Elementary Theory of Eisenstein Series

Elementary Theory of Eisenstein Series

by
Tomio KUBOTA
Professor of Mathematics
University of Maryland, U.S.A
and
Nagoya University, Japan

A HALSTED PRESS BOOK

KODANSHA LTD.
Tokyo
JOHN WILEY & SONS
New York-London-Sydney-Toronto

KODANSHA SCIENTIFIC BOOKS

Library of Congress Cataloging in Publication Data

Kubota, Tomio.
Elementary theory of Eisenstein series.
(Kodansha scientific books)
"Halsted Press book."
Includes bibliographical references.
1. Eisenstein series. 2. Analytic continuation.
3. Representations of groups. I. Title.
QA331.K83 515′.9 73-1687
ISBN 0-470-50920-1

KODANSHA EDP NO.: 3041-247416-2253(0)

Published in Japan
by
KODANSHA LTD.
12-21, Otowa 2-chome, Bunkyo-ku, Tokyo 112, Japan
Published
by
HALSTED PRESS
a Division of John Wiley & Sons, Inc.
605 Third Avenue, New York, N.Y. 10016, U.S.A.

PRINTED IN JAPAN

Contents

Preface

The contents of this book are based principally on a course given by the author in 1970–71 at the University of Maryland. Since it was not an express aim of the course to develop a full and complete theory, several important aspects of real analytic Eisenstein series are not included in the present text. The main part of the book gives an expository description of the analytic continuation of Eisenstein series concerning $SL(2, \boldsymbol{R})$. In addition, spectral decomposition of the space of square integrable automorphic functions with respect to a general Fuchsian group is outlined, since it constitutes one of the most interesting applications of real analytic Eisenstein series. However, the primary aim of the book is to summarize the characteristic properties of real analytic Eisenstein series; consequently, items such as purely functional analytic arguments are left in rather abbreviated form.

The author is grateful to Mr. Isamu Kobayashi and Mr. Yoshiyuki Kitaoka, who kindly helped in preparing and proof-reading the manuscript. The author furthermore wishes to express his sincere thanks to Professor Larry J. Goldstein, who participated not only in a scientific but also in a rhetoric check of the manuscript, and to Professor Walter L. Baily, Jr. for his final, critical reading of the text prior to publication.

December, 1972 Tomio KUBOTA

Introduction

An Eisenstein series $E(z, s)$, for the purposes of this book, is a function of two variables z and s, where z denotes a point in the complex upper half plane $H = \{z \in \boldsymbol{C} \mid \operatorname{Im} z > 0\}$, and s is a complex variable. The function $E(z, s)$ is not complex analytic with respect to z, but real analytic in the sense that $E(z, s)$ satisfies the differential equation $y^2\left(\frac{\partial^2}{\partial x^2} + \frac{\partial^2}{\partial y^2}\right)E = \lambda E$, $(z = x + \sqrt{-1}y,\ \lambda \in \boldsymbol{C})$, while $E(z, s)$ is meromorphic with respect to s. Moreover, $E(z, s)$ is an automorphic function with respect to z, that is, $E(z, s)$ is invariant under a certain group Γ of fractional linear transformations $z \to \sigma z = \frac{az + b}{cz + d}$, $(a,b,c,d \in \boldsymbol{R},\ ad - bc = 1)$; in other words, $E(\sigma z, s) = E(z, s)$ whenever $\sigma \in \Gamma$.

A typical example of an Eisenstein series is given by

$$E(z, s) = \frac{1}{2}\sum \frac{y^s}{|cz + d|^{2s}}, \qquad (c, d \in \boldsymbol{Z},\ (c, d) = 1).$$

In fact, if we denote by Γ the elliptic modular group $SL(2, \boldsymbol{Z})$, and by $y(z)$ the imaginary part of $z = x + \sqrt{-1}y \in \boldsymbol{C}$, then the above series is equal to $\Sigma y(\sigma z)^s$, where the sum ranges over all the left cosets in the quotient $\Gamma_\infty\backslash\Gamma$ with respect to the group Γ_∞ of all triangular matrices $\begin{pmatrix} a & b \\ 0 & d \end{pmatrix} \in \Gamma$, $(a = d = \pm 1)$. Since $y(z)$ is invariant under Γ_∞, the sum is well defined and is automorphic with respect to Γ. Since, on the other hand, y^s satisfies $y^2\left(\frac{\partial^2}{\partial x^2} + \frac{\partial^2}{\partial y^2}\right)y^s = s(s-1)y^s$, and since $D = y^2\left(\frac{\partial^2}{\partial x^2} + \frac{\partial^2}{\partial y^2}\right)$ is an invariant differential operator in the sense that we have $D(f(\sigma z)) = (Df)(\sigma z)$ for every $\sigma \in SL(2, \boldsymbol{R})$, we see at once that $DE(z, s) = \lambda E(z, s)$ holds with $\lambda = s(s - 1)$.

The functions defined by these series will, however, not have any meaning unless the series converge. As a matter of fact, the series

$\Sigma y(\sigma z)^s$ converges if and only if Re $s > 1$, so that the situation becomes quite similar to that for Dirichlet series, and we must consider the problem of analytic continuation of $E(z, s)$ with respect to s. This is the starting point of the theory of real analytic Eisenstein series. The main result which we obtain in this direction is that every Eisenstein series has an analytic continuation onto the whole s-plane and satisfies a functional equation, that is, a certain equality relating $E(z, s)$ and $E(z, 1 - s)$, exactly as in the case of many zeta functions. Because of this fact Eisenstein series are applicable to the study of various Dirichlet series appearing in number theory, if, for example, we consider an arithmetically defined discrete subgroup of $SL(2, \boldsymbol{R})$ such as the elliptic modular group. Another important application of Eisenstein series can be found in the theory of unitary representations of Lie groups. The Eisenstein series considered in this book are mainly related to the group $SL(2, \boldsymbol{R})$. We shall study some of the facts concerning representations of $SL(2, \boldsymbol{R})$ which derive from an investigation of the values of $E(z, \frac{1}{2} + \sqrt{-1}t)$, $(t \in \boldsymbol{R})$.

The theory of real analytic Eisenstein series was introduced in the work of A. Selberg (ref. 1), although proofs were not given in a detailed form. Several years after the work of Selberg (see also ref. 2), Langlands generalized the theory to the case of more general Lie groups. This theory of course includes, with proofs, the original Selberg theory as a special case, but there are unfortunately very few references available for Langlands' theory, except the article (ref. 3). Now, Harish-Chandra has published a text book (ref. 4) for Selberg-Langlands' theory in the form of lecture notes. For beginners, however, these lecture notes might be difficult, because they are at a comparatively high level. On the other hand, there are many papers in representation theory related to various aspects of Selberg's theory: for example, Godement's papers (ref. 5–6), and many papers of the Russian school (ref. 7–9, among others). These are inspiring works; for example, Godement's papers contain, at least for the elliptic modular group, a complete and very understandable explanation of the relationship between representation theory and Eisenstein series, although Selberg's theory itself is not developed there, nor in the Russian papers.

The aim of this book is to give a completely expository introduction of Selberg's theory as well as of its application to representation theory. We shall treat only the case of $SL(2, \boldsymbol{R})$ and its discrete subgroups. However, we do not restrict ourselves only to the case of the modular group, so that the method used in this book is applicable, in principle, to much more general situations.

The theory of Eisenstein series is a basic tool required in the study of the arithmetic of discontinuous groups. At the same time the theory is so interesting and beautiful that it is worthy of wider popularization.

References

1. A. Selberg, *Harmonic analysis and discontinuous groups in weakly symmetric Riemannian spaces with applications to Dirichlet series. J. Indian Math. Soc.*, **20**, 47–87 (1956).
2. A Selberg, *Discontinuous groups and harmonic analysis*, Proc. Int. Congr. Math., Stockholm, p. 177–189, 1962.
3. R. P. Langlands, *Eisenstein series*, Proc. Symp. Pure Math., A.M.S., **9**, 235–252 (1966).
4. Harish-Chandra, *Automorphic forms on semisimple Lie groups. Lecture Notes in Math.*, 62, Springer, 1968.
5. R. Godement, *The decomposition of* $L^2(G\backslash\Gamma)$ *for* $\Gamma = SL(2, \boldsymbol{Z})$. Proc. Symp. Pure Math., A.M.S., **9**, 211–224 (1966).
6. R. Godement, *The spectral decomposition of cusp forms. Ibid.*, 225–234.
7. I. M. Gel'fand and I. I. Pyatetskii-Shapiro, *Representation theory and theory of automorphic functions. Am. Math. Soc. Transl.*, ser. 2, **26**, 173–200 (1963).
8. I. M. Gel'fand, M. I. Graev and N. Ya. Vilenkin, *Integral geometry and representation theory* (*Generalized functions*, vol. 5), Moscow, 1962. Translation: Academic Press, 1966.
9. I. M. Gel'fand, M. I. Graev and I. I. Pyatetskii-Shapiro, *Representation theory and automorphic functions* (*Generalized functions*, vol. 6), Moscow, 1966. Translation: Saunders, 1969.

CHAPTER I

Preliminaries

§1.1. The upper half plane

We denote by H the complex *upper half plane* $\{z = x + \sqrt{-1}y \in \boldsymbol{C} \mid \operatorname{Re} z = y > 0\}$. An element $\sigma = \begin{pmatrix} a & b \\ c & d \end{pmatrix} \in SL(2, \boldsymbol{R})$ operates on H by means of the fractional *linear transformation* $z \to \sigma z = \dfrac{az + b}{cz + d}$. The transformation σ maps H onto H, and at the same time σ maps the lower half plane $\bar{H}$ on $\bar{H}$, and $\boldsymbol{R} \cup \{\infty\}$ on itself. Thus, the complex plane $\boldsymbol{C}$ is divided into three orbits of $SL(2, \boldsymbol{R})$.

We now view the upper half plane H from a more group-theoretical standpoint. Let K be the *maximal compact subgroup* of $G = SL(2, \boldsymbol{R})$ consisting of all matrices of the form $\begin{pmatrix} \cos\theta & -\sin\theta \\ \sin\theta & \cos\theta \end{pmatrix}$, $\theta \in \boldsymbol{R}$; then, G/K is identified with H by the mapping $G \ni \sigma \to z = \sigma\sqrt{-1} \in H$. In other words, if we denote by N, (resp. A), the group of all matrices of the form $\begin{pmatrix} 1 & x \\ & 1 \end{pmatrix}$, (resp. $\begin{pmatrix} \sqrt{y} & \\ & \sqrt{y}^{-1} \end{pmatrix}$), $(x \in \boldsymbol{R},\ y > 0)$, and consider the *Iwasawa decomposition* $G = NAK$ given by the unique decomposition of $\sigma \in G$ in the form $\sigma = n(x)a(y)k(\theta)$ with

$$(1.1.1) \qquad n(x) = \begin{pmatrix} 1 & x \\ & 1 \end{pmatrix}, \quad a(y) = \begin{pmatrix} \sqrt{y} & \\ & \sqrt{y}^{-1} \end{pmatrix}, \quad k(\theta) = \begin{pmatrix} \cos\theta & -\sin\theta \\ \sin\theta & \cos\theta \end{pmatrix},$$

then we have $z = x + \sqrt{-1}y$.

To the upper half plane H, we give now a Riemannian structure determined by

$$(1.1.2) \qquad ds^2 = \frac{1}{y^2}(dx^2 + dy^2)\,.$$

This is (up to scalar multiples) the unique, *G-invariant Riemannian metric* on H. The geodesics are arcs (half circles) orthogonal to the real axis, and

H becomes a non-euclidean Riemannian space. The metric (1.1.2) gives rise naturally to a *G-invariant measure* on H whose explicit form is

$$dz = \frac{1}{y^2}\,dxdy\,. \tag{1.1.3}$$

§1.2. Discontinuous groups

A *discontinuous group* is a discrete subgroup of the group $PSL(2, \boldsymbol{R}) = G/\{\pm 1\}$. Since the elements ± 1 in the center of $SL(2, \boldsymbol{R})$ are the only elements in $SL(2, \boldsymbol{R})$ inducing the identity mapping on H, $G/\{\pm 1\}$ is exactly the group of fractional linear transformations. However, if no confusion is possible, we shall denote a linear transformation by a matrix in G representing it. In particular, a discontinuous group Γ, which is a group of fractional linear transformations, will also sometimes be handled as if it were a matric group. In fact, in order that a subgroup Γ of $PSL(2, \boldsymbol{R})$ be a discontinuous group, it is necessary and sufficient that the inverse image of Γ in G be discrete.

A *fundamental domain* $\mathscr{D}$ of a discontinuous group Γ is an open domain in the upper half plane H with the following properties: i) if σ and τ are two different elements in Γ, then $\sigma\mathscr{D} \cap \tau\mathscr{D} = \phi$; ii) the closure of $\cup\, \sigma\mathscr{D}$, $(\sigma \in \Gamma)$, coincides with H.

Two points z_1, z_2 on $\boldsymbol{C}$ will be called *equivalent* with respect to Γ, if there exists a $\sigma \in \Gamma$ such that $\sigma z_1 = z_2$. Therefore, a fundamental domain $\mathscr{D}$ of Γ means a domain such that two inner points of $\mathscr{D}$ are not equivalent, and such that any point in H is equivalent to a point on the closure of $\mathscr{D}$. The domain $\mathscr{D}$ will sometimes be denoted symbolically by $\Gamma\backslash H$.

For any discontinuous group Γ, a fundamental domain always exists. Actually, there exists a fundamental domain which is a non-euclidean polygon in H. To show this, take a point $z_0 \in H$ which is not a fixed point of any $\sigma \in \Gamma$ different from the identity, and consider the set $\mathscr{D}$ of points z in H such that the non-euclidean distance $d(z, z_0)$ between z and z_0 is smaller than any other $d(z, \sigma z_0)$, $(\sigma \neq 1, \sigma \in \Gamma)$. This $\mathscr{D}$ is then a fundamental domain for Γ. Such a domain is called a *normal polygon*.

For more precise statements on the fundamental domain, we have to recall here the classification of linear transformations. Let σ be a linear transformation which is not the identity; then as the Jordan canonical form of σ, we can take one of the following:

i) $\begin{pmatrix} \varepsilon & \\ & \bar{\varepsilon} \end{pmatrix}$, $\varepsilon \in \boldsymbol{C}$, $|\varepsilon| = 1$,

ii) $\begin{pmatrix} t & \\ & t^{-1} \end{pmatrix}$, $t \in \boldsymbol{R}$,

$$\text{iii)} \quad \begin{pmatrix} 1 & x \\ & 1 \end{pmatrix}, \quad x \in \boldsymbol{R}.$$

The transformation σ is called *elliptic, hyperbolic*, resp. *parabolic*, according as the Jordan canonical form of σ, as a matrix, is of type i), ii), resp. iii).

The invariant (non-euclidean) measure of the fundamental domain $\mathscr{D}$ of a discontinuous group Γ is not always finite, and, if the measure of $\Gamma\backslash H$ is finite, then Γ is called a discontinuous group of *finite type*. In this book, we shall deal exclusively with discontinuous groups of finite type. A discontinuous group Γ of finite type has in fact many convenient properties. First of all, Γ is finitely generated and is defined by finitely many relations among its generators; this is a theorem of Siegel.† A normal polygon of a finitely generated discontinuous group Γ has a finite number of sides and vertices. But, if a side in the usual sense contains a fixed point of an elliptic element of order two in Γ, then the side should be regarded as consisting of two different sides and the fixed point should be regarded as a vertex joining the two sides. According to this convention, the set of sides of $\mathscr{D}$ is divided into sets of equivalent pairs of sides in such a way that one side in an equivalent pair is mapped on the other by an element $\gamma_i \in \Gamma$ which maps $\mathscr{D}$ on a fundamental domain adjacent to $\mathscr{D}$. The finite set of elements γ_i forms a set of generators of Γ. It is also possible to give a finite, generating set of relations for Γ in a similar geometric way. This will not be discussed here, but it should be noted that the geometric investigation of the structure of Γ is based upon the fact that Γ is the fundamental group, possibly with ramification, of the surface built up from $\mathscr{D}$ by identifying the two sides in each equivalent pair.

To proceed further, we now have to introduce the notion of cusp. Let κ be a real number or ∞, and let Γ_κ be the *stabilizer* of κ in Γ, i.e. $\Gamma_\kappa = \{\sigma \in \Gamma \mid \sigma\kappa = \kappa\}$. We then call κ a *cusp* of Γ, if Γ_κ is generated by a parabolic element. If κ is a cusp, then we can take a non-euclidean strip as a fundamental domain of Γ_κ. If $\kappa = \infty$ is a cusp, then the strip becomes an euclidean strip in the usual sense. In order that Γ be of finite type, it is necessary and sufficient that any point of $\boldsymbol{R} \cup \{\infty\}$ which lies on the boundary of a fundamental domain of Γ be a cusp of Γ. Inequivalent cusps are in one to one correspondence with conjugacy classes of maximal subgroups of Γ having a parabolic element as generator. If Γ is of finite type, then the number of inequivalent cusps is finite. Proofs of these facts will, however, be omitted in this book, because it is

† See e.g. J. Lehner, *A short course in automorphic functions*, Holt, Rinehart and Winston, 1966.

sufficient to start our investigation with a discontinuous group of finite type for which a normal polygon has a finite number of sides and vertices, and has a finite number of cusps as its only boundary points in $\boldsymbol{R} \cup \{\infty\}$. Examples of the shape of a fundamental domain of a discontinuous group are given by the following figures:

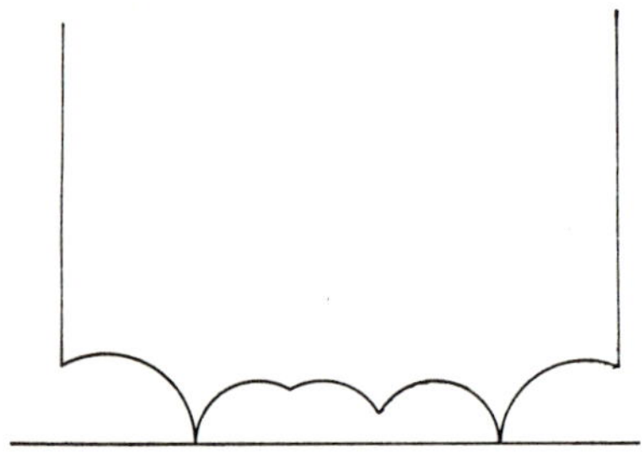

Fig. 1

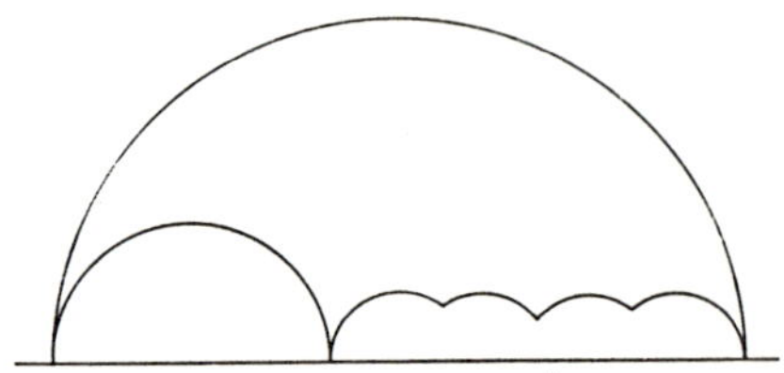

Fig. 2

Fig. 1 indicates a fundamental domain for which the point ∞ is a cusp on the boundary, while ∞ is not on the boundary of the fundamental domain in Fig. 2. If Γ has no cusp, a fundamental domain of Γ is illustrated by Fig. 3. Such a fundamental domain is called *compact*.

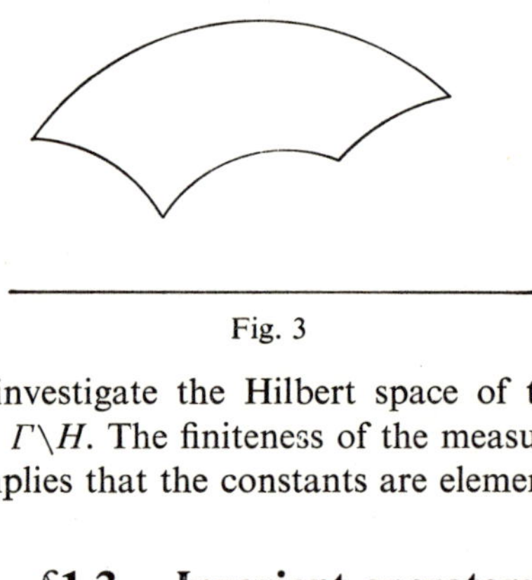

Fig. 3

We shall later investigate the Hilbert space of the square integrable functions on $\mathscr{D} = \Gamma\backslash H$. The finiteness of the measure of $\mathscr{D}$ is important here, because it implies that the constants are elements of $L^2(\Gamma\backslash H)$.

§1.3. Invariant operators

By a *linear operator*, we mean a linear operator on the space of complex valued functions on H. The operators which are investigated in this book are, in the main, either differential operators or *integral operators* defined by

$$\int_H k(z, z')f(z')dz' = (Lf)(z) \tag{1.3.1}$$

with a *kernel function* $k(z, z')$.

For $\sigma \in G$, the mapping $f(z) \to f(\sigma z)$ also defines a linear operator. This will be denoted by T_σ. A linear operator L is called an *invariant operator*, if it commutes with all T_σ, $(\sigma \in G)$, i.e., if we have $L(f(\sigma z)) = (Lf)(\sigma z)$. In order that a kernel function $k(z, z')$ define an invariant operator, it is necessary and sufficient that $k(\sigma z, \sigma z') = k(z, z')$ hold for every $\sigma \in G$.

There is another convenient and important linear operator for our investigation, called a mean value operator. Let z_0 be a point on H, and let $\sigma_0 \in G$ be a transformation such that $\sigma_0 z_0 = \sqrt{-1}$. Then the stabilizer in G of z_0 is $\sigma_0^{-1}K\sigma_0$. Let $k(\theta)$ be as in (1.1.1); then the *mean value operator* M_{z_0} around z_0 is defined by

$$(M_{z_0}f)(z) = \frac{1}{2\pi}\int_0^{2\pi} f(\sigma_0^{-1}k(\theta)\sigma_0 z)d\theta \,.$$

The mean value operator can be applied not only to functions but also to differential forms and linear operators. Here, the operation $L \to L^\sigma$ of $\sigma \in G$ on an operator L should be defined by $L^\sigma = T_\sigma L T_\sigma^{-1}$, so that the image of L of the mean value operator M_{z_0} is the mean

$$\frac{1}{2\pi}\int_0^{2\pi} L^\sigma d\theta$$

with $\sigma = \sigma_0^{-1}k(\theta)\sigma_0 \in \sigma_0^{-1}K\sigma_0$. Using this, it is easy to see that there is no non-zero differential operator of order 1 at the point $\sqrt{-1}$ which is invariant under the mean value operator around $\sqrt{-1}$. On the other hand, invariant differential operators are in one to one correspondence with their restrictions to one point, because a necessary and sufficient condition for an operator L to be invariant is that the restrictions L_1, L_2 of L to any two points $z_1, z_2 \in H$ satisfy $T_\sigma L_1 T_\sigma^{-1} = L_2$ whenever $\sigma z_2 = z_1$. Therefore, in particular, there is no G-invariant differential operator of order 1 on H. Similarly, it may be shown that every differential operator at $\sqrt{-1}$ invariant under the mean value operator around $\sqrt{-1}$ must be a polynomial in $\dfrac{\partial^2}{\partial x^2} + \dfrac{\partial^2}{\partial y^2}$, because a polynomial of two variables X and Y invariant under all the orthogonal transformations in the two variables X and Y must be a polynomial in $X^2 + Y^2$. Thus, we can conclude that

$$D = y^2\left(\frac{\partial^2}{\partial x^2} + \frac{\partial^2}{\partial y^2}\right) \tag{1.3.2}$$

is the G-invariant differential operator on H of lowest order, and that every G-invariant differential operator on H is a polynomial in D with constant coefficients. This kind of result is contained in the general framework of differential geometry, and D in (1.3.2) is the *Laplacian* of the Riemannian space H.

We now propose to investigate more precisely the invariant operators on H. For this purpose, we have first to consider the functions $k(z, z')$ on $H \times H$ satisfying $k(\sigma z, \sigma z') = k(z, z')$ for every $\sigma \in G$. Such a function defines an invariant integral operator by means of (1.3.1), and is called a *point pair invariant*. Since there is a natural projection $G \to H = G/K$, a function on H may always be regarded as a function on G. In particular, a point pair invariant $k(z, z')$ determines a function $k(g, g')$ on $G \times G$. It is obvious then that there exists a function F on G such that $F(g'^{-1} g) = k(g, g')$, and this function F has a particular property, namely, $F(kgk') = F(g)$ for any $k, k' \in K$. In other words, F is constant on every double coset KgK. Conversely, any function F on G which is constant on every double coset KgK will define a point pair invariant on H by $k(g, g') = F(g'^{-1} g)$. If f is a function on H regarded at the same time as a function on G, then we have

$$\int_H k(z, z')f(z')dz' = \int_G f(g')F(g'^{-1} g)dg' = \int_G f(gg')F(g'^{-1})dg',$$

where dg' is the Haar measure on G such that the measure of K is normalized to be 1. This means that the operation of the integral operator determined by k is nothing else than the mapping $f \to f * F$, where $*$ denotes the convolution on G.

A more important result can be derived if we use the so-called Cartan decomposition. For an element $g \in G$, there exist $k(\theta), k(\theta') \in K$ and $a(t) \in A$ such that $g = k(\theta)a(t)k(\theta')$, where $k(\theta)$, $k(\theta')$, and $a(t)$ are defined as in the Iwasawa decomposition, but the numbers θ and t appearing here are different from those in the Iwasawa decomposition. The decomposition $G = KAK$ given by the above formula is called the *Cartan decomposition.* The elements $k(\theta)$, $k(\theta')$ in the Cartan decomposition are unique up to ± 1. The assertion of the Cartan decomposition is essentially the same thing as the statement that a point pair invariant $k(z, z')$ is a function of the non-euclidean distance $d(z, z')$ between z and z'. Since the inverse σ^{-1} of $\sigma = \begin{pmatrix} a & b \\ c & d \end{pmatrix} \in G$ is given by $\sigma^{-1} = \begin{pmatrix} d & -b \\ -c & a \end{pmatrix}$, we see that $KgK = Kg^{-1}K$ for any $g \in G$. This implies the important fact that for a function F on G which is constant on every double coset KgK, we have $F(g) = F(g^{-1})$.

Using these results, we can now prove a theorem on the commutativity of the ring of invariant operators. The analytic side of the theorem will be treated somewhat sketchily, since the essential part of the proof of the theorem consists of rather formal and algebraic arguments, excepting possibly the commutativity of an invariant operator with a mean value operator. Such commutativity, however, always holds for the operators and functions which we use in this book.

THEOREM 1.3.1. Two invariant operators L and L' on H commute with each other.

Proof. We take to begin with a point pair invariant $k(z,z')$ as a function of the first variable z, and will prove that $LL'k = L'Lk$. To do this, choose a function F on G such that $k(z, z') = F(g'^{-1}g)$. Then, since $F(g) = F(g^{-1})$ and since L is invariant, we have

$$\begin{aligned} L_g(F(g'^{-1}g)) &= (LF)(g'^{-1}g) = (LF)(g^{-1}g') \\ &= L_{g'}(F(g^{-1}g')) = L_{g'}(F(g'^{-1}g)), \end{aligned}$$

where L_g, say, means that L operates on g. Thus, we see that

$$L_z k(z, z') = L_{z'} k(z, z'),$$

and the same formula holds for L', too. This implies $LL'k = L'Lk$, because any two operators operating on different variables are commutative.

Suppose now that $f(z)$ is an arbitrary function on H, and let $M_{z'}$ be the mean value operator around z'. Then, there exists a point pair invariant $k_{z'}$ depending on z' such that $M_{z'}f(z) = k_{z'}(z, z')$. So, $LL'M_{z'}f(z) = L'LM_{z'}f(z)$ for any z'. Since L, L' are invariant, $LL'M_{z'} = L'LM_{z'}$ entails $M_{z'}LL' = M_{z'}L'L$. But, $M_{z'}$ does not alter the value of any function at z', and z' is arbitrary. Hence, putting $z' = z$, we have $LL' = L'L$. *q.e.d.*

The commutativity in this theorem holds for a wider class of groups for which there exists an anti-involution $g \to g^*$ satisfying $KgK = Kg^*K$, K being a compact subgroup.

To get another important theorem on invariant operators, we have to introduce the notion of zonal spherical functions. A *zonal spherical function* on H belonging to the eigenvalue λ is a function ω on H satisfying the following conditions:

i) $\omega(\sqrt{-1}) = 1$,
ii) $\omega(k(\theta)z) = \omega(z)$ for every $k(\theta)$ in (1.1.1),
iii) $D\omega = \lambda\omega$ for the Laplacian D in (1.3.2).

The function $\omega(z)$ can be viewed as a function on G which is constant on each double coset KgK, $(g \in G)$.

A zonal spherical function $\omega(z)$ is uniquely determined by λ. To show this, let L be a differential operator at the point $\sqrt{-1}$, and put $\sigma =$

$k(\theta)$. Then, since $L^\sigma = T_\sigma L T_\sigma^{-1} = L T_\sigma^{-1}$ is well-defined and $L^\sigma\omega = L\omega$, the value of $L\omega$ is determined if we know the value $L_0\omega$ for the operator L_0 derived from L by applying the mean value operator at $z = \sqrt{-1}$. Since furthermore L_0 is obtained by the restriction to the point $z = \sqrt{-1}$ of an invariant differential operator, $L\omega$ is determined whenever we know the value at $z = \sqrt{-1}$ of $L'\omega$ for all invariant differential operators L'. This implies the uniqueness of ω, because every invariant differential operator is a polynomial in D, and because ω is a real analytic function of x and y.

To prove the existence of ω, we take simply the function y^s, and apply the mean value operator around $\sqrt{-1}$.

We can now prove an important theorem that an eigenfunction of the Laplacian is an eigenfunction of all invariant integral operators. This theorem is basic in the theory of harmonic functions.

THEOREM 1.3.2. Let $f(z)$ be a function on H satisfying $Df = \lambda f$, where D is the Laplacian and λ is a complex number, and let L be an invariant operator. Then, there exists a constant $\Lambda_L(\lambda)$ depending only on L and λ, and not on the individual function f, such that $Lf = \Lambda_L(\lambda)f$.

Proof. Let ω be the zonal spherical function belonging to the eigenvalue λ, and consider the point pair invariant defined by $\omega(z, z') = \omega(g'^{-1} g)$, $(g, g' \in G)$. Then, as a function of z, $\omega(z, z')$ is invariant under every element in the stabilizer K' in G of z', and $D\omega = \lambda\omega$. Since the ring of invariant operators is commutative by Theorem 1.3.1, we see that the function $\omega_1(z) = L_z\omega(z, z')$ is also, as a function of z, invariant under K' and $D\omega_1 = \lambda\omega_1$, where L_z means that L operates on the first variable z. Therefore, the uniqueness of the zonal spherical function assures that we have $L_z\omega(z, z') = \Lambda_L(\lambda)\omega(z, z')$ with a constant $\Lambda_L(\lambda)$ depending on L and λ.

Let now $f(z)$ be an arbitrary function satisfying $DF = \lambda f$, and let $M_{z'}$ be the mean value operator around $z' \in H$. Then, since $M_{z'}f(z)$ coincides with the above $\omega(g'^{-1} g)$ up to a constant factor, we see that $LM_{z'}f(z) = \Lambda_L(\lambda)M_{z'}f(z)$. From this it follows that $M_{z'}Lf(z) = M_{z'}\Lambda_L(\lambda)f(z)$, and, putting $z' = z$, we have $Lf(z) = \Lambda_L(\lambda)f(z)$. *q.e.d.*

In this theorem also, analytic conditions such as the commutativity of an invariant operator and a mean value operator are tacitly assumed; the situation is similar to that in Theorem 1.3.1.

It should be noted that the converse of Theorem 1.3.2 is also true. In fact, we have

THEOREM 1.3.3. Let $f(z)$ be a function on H which is an eigenfunction of every invariant integral operator determined by a point pair invariant $k(z, z')$ with the property that the corresponding function $F(g)$ defined as in the proof of Theorem 1.3.1 is C^∞ and has compact support. Then, $f(z)$ is an eigenfunction of the Laplacian on H.

Proof. Let $k(z, z')$ be a point pair invariant with the property stated in the theorem, and assume that

$$\int_H k(z, z')f(z')dz' = \Lambda f(z)$$

with $\Lambda \neq 0$. Such a k always exists unless $f(z) = 0$ almost everywhere. Moreover, we may assume here, by appropriate choice of k, that $D_z k(z, z') = k_1(z, z')$ exists. Since k_1 is again a point pair invariant, we have

$$\Lambda Df(z) = \int_H k_1(z, z')f(z')dz' = \Lambda_1 f(z).$$

This yields $Df = \Lambda_1 \Lambda^{-1} f$. *q.e.d.*

The limit of a series of eigenfunctions of the Laplacian can be treated more easily than that of a series of continuous or differentiable functions. The situation is similar to that for the limit of a series of holomorphic functions, which is fairly easy to investigate, because there are convenient integral representations. In this connection, we state here a theorem which will be useful in obtaining the analytic continuation of Eisenstein series.

THEOREM 1.3.4. Let $\mathfrak{H}$ be the space of square integrable functions on a domain S in the upper half plane H, where the inner product of two functions f_1, f_2 is defined by

$$(f_1, f_2) = \int_S f_1(z)\overline{f_2(z)}dz.$$

Suppose that $g_1, g_2, \ldots$ is a weakly convergent sequence of functions of bounded norm in $\mathfrak{H}$ such that $Dg_i = \lambda_i g_i$, where $\lambda_i \in \boldsymbol{C}$ and $\lim \lambda_i = \lambda$. Then, $\{g_i\}$ converges pointwise to a function $g \in \mathfrak{H}$ such that $Dg = \lambda g$.

Proof. Let z_0 be a point in S and assume the disc with center z_0 and with radius ε lies in S. Let $k(z, z')$ be a C^∞ point pair invariant such that $k(z, z') = 0$ if the distance $d(z, z')$ between z and z' is bigger than ε. Then, it follows from Theorem 1.3.2 that

$$\int_H k(z_0, z')g_i(z')dz' = \Lambda_i g_i(z_0) \tag{1.3.3}$$

for all i, and, by considering the operation of the integral operator on

y^s, $(s \in C)$, one may see that $\lim \Lambda_i = \Lambda$ exists. Furthermore, it is possible to choose k such that $\Lambda \neq 0$. Let now g_0 be a function in $\mathfrak{H}$ to which g_i converges weakly. Then, by the definition of weak convergence, $\lim g_i(z_0) = g(z_0)$, if we define $g(z)$ by

$$\int_H k(z, z')g_0(z')dz' = \Lambda g(z).$$

The function $g(z)$ is defined on a certain subdomain S' of S containing z_0. Since, however, $z_0 \in S$ is arbitrary, $g(z)$ can be regarded as a function defined on S to which $g_i(z)$ converges pointwise. Now, Fatou's theorem shows that $g(z)$ belongs to $\mathfrak{H}$ as it is the pointwise limit of functions $g_i(z)$ of bounded norm, and at the same time (1.3.3) implies

$$\int_H k(z, z')g(z')dz' = \Lambda g(z), \qquad (z \in S'),$$

and

$$\int_H k_1(z, z')g(z')dz' = \Lambda_1 g(z), \qquad (z \in S'),$$

with $k_1(z, z') = D_z k(z, z')$ and $D_z = D$ operating on z. Hence, the arguments in the proof of Theorem 1.3.3 show that $Dg = \Lambda^{-1}\Lambda_1 g$, i.e., g is an eigenfunction of D. Since $\Lambda = \lim \Lambda_i$ and $\Lambda_1 = \lim \Lambda_{i,1}$ with

$$\int_H k_1(z, z')g_i(z')dz' = \Lambda_{i,1} g_i(z),$$

we have $\Lambda^{-1}\Lambda_1 = \lim \Lambda_i^{-1}\Lambda_{i,1} = \lim \lambda_i$, and consequently $Dg = \lambda g$.

q.e.d.

CHAPTER II

Eisenstein series in the domain of convergence

§2.1. Definition of Eisenstein series

Let Γ be a discontinuous group of finite type operating on the upper half plane H, and let $\kappa_1, \kappa_2, \ldots, \kappa_h$ be a maximal set of cusps of Γ which are not equivalent with respect to Γ. Such a set will be called a *complete set of inequivalent cusps*. Let Γ_i be the stabilizer in Γ of κ_i, i.e. $\Gamma_i = \{\sigma \in \Gamma \mid \sigma\kappa_i = \kappa_i\}$, and fix an element $\sigma_i \in G = SL(2, \boldsymbol{R})$ such that $\sigma_i\infty = \kappa_i$ and such that $\sigma_i^{-1}\Gamma_i\sigma_i$ is equal to the group Γ_0 of all matrices of the form $\begin{pmatrix} 1 & b \\ & 1 \end{pmatrix}$ with $b \in \boldsymbol{Z}$. Furthermore, denote by $y(z)$ the imaginary part of $z \in H$. The *Eisenstein series* $E_i(z, s)$ for the cusp κ_i is then defined by

$$E_i(z, s) = \sum y(\sigma_i^{-1}\sigma z)^s, \qquad (\sigma \in \Gamma_i\backslash\Gamma), \tag{2.1.1}$$

where s is a complex variable.

From this definition, it follows immediately that, whenever the series converges uniformly, we have $E_i(\sigma z, s) = E_i(z, s)$ for any $\sigma \in \Gamma$, and $DE_i = \lambda E_i$ with $\lambda = s(s-1)$, where D denotes as in (1.3.2) the Laplacian of H. Furthermore, we see that the transformation σ_i is, up to the multiplication by a translation $\begin{pmatrix} 1 & x \\ & 1 \end{pmatrix}$ from the right, uniquely determined for a cusp x_i, and that E_i does not depend on the particular choice of a cusp x_i among equivalent ones.

In this book, we call a discontinuous group Γ of finite type *reduced at* ∞, if ∞ is a cusp, and if the stabilizer $\Gamma_\infty = \{\sigma \in \Gamma \mid \sigma\infty = \infty\}$ of ∞ is equal to $\Gamma_0 = \{\begin{pmatrix} 1 & b \\ & 1 \end{pmatrix} \mid b \in \boldsymbol{Z}\}$. At many places in the theory of Eisenstein series, it is sufficient to consider a group Γ reduced at ∞. If κ_i is a cusp of a discontinuous group Γ of finite type, and if σ_i is as above, then the group $\sigma_i^{-1}\Gamma\sigma_i$ is reduced at ∞. The elliptic modular group is an example of a discontinuous group of finite type reduced at ∞.

If a discontinuous group Γ has no cusp, i.e., if the fundamental domain of Γ is compact, then there are no Eisenstein series. So, we shall consider in this book only those discontinuous groups which have a non-compact fundamental domain.

THEOREM 2.1.1. An Eisenstein series converges absolutely, if Re $s > 1$.

Proof. We may assume that s is a positive, real number, and shall assume that Γ is reduced at ∞, which is also no restriction of generality. Furthermore, it is sufficient to prove the theorem for the Eisenstein series E_1 for $\kappa_1 = \infty$ which we choose as the first among a set of inequivalent cusps $\kappa_1, \kappa_2, \ldots, \kappa_h$. For the sake of simplicity, we denote E_1 simply by E, i.e., $E(z, s) = \sum y(\sigma z)^s$, $(\sigma \in \Gamma_0\backslash\Gamma)$. Now let S be the strip determined by $0 < \text{Re}\, s < 1$. Then, for any fixed $z_0 \in S$, we may take as a set of orbit representatives $\{\sigma z_0 \mid \sigma \in \Gamma_0\backslash\Gamma\}$ modulo Γ_0, the set M of all points in S equivalent to z_0. Since ∞ is a cusp of Γ, the set of numbers $y(\sigma z_0)$, $(\sigma \in \Gamma_0\backslash\Gamma)$, is bounded.

Consider now the point pair invariant $k(z, z')$ defined by $k(z, z') = 1$ or 0 according as the distance $d(z, z')$ between z and z' is smaller than a positive number ε, or not. Then, since every $y(\sigma z)^s$ is an eigenfunction of D, it is also an eigenfunction of the invariant integral operator defined by $k(z, z')$. More specifically, by Theorem 1.3.2, there exist a positive number Λ_ε, depending on ε and not on σ, such that

$$\int_H k(z_0, z')y(\sigma z')^s dz' = \Lambda_\varepsilon y(\sigma z_0)^s.$$

This integral is, however, the integral of y^s in a non-euclidean circle with radius ε and with center σz_0. Therefore, taking ε sufficiently small, we can conclude that the Eisenstein series converges at z_0 whenever the integral $\int_0^Y \int_0^1 y^s \frac{dxdy}{y^2}$ exists, where Y denotes a suitable positive constant. This proves the theorem. *q.e.d.*

This proof shows uniform convergence on compact sets, too. The proof of the converse of this theorem will be given in §4.4.

THEOREM 2.1.2. Let Γ be a discontinuous group of finite type reduced at infinity, and let $E(z, s)$ be the Eisenstein series for the cusp ∞ of Γ. Then, provided that Re $s > 1$, $E(z, s) - y^s$ is bounded when $y \to \infty$. Furthermore, $E(z, s)$ is bounded when z tends to a cusp which is not equivalent to ∞.

Proof. Notation being as in the proof of Theorem 2.1.1, $E(z, s) - y^s$ is equal to $\sum y(\sigma z)^s$, $(\sigma \in \Gamma_0 \backslash \Gamma,\ \sigma \neq 1)$. Therefore, replacing the sum by an integral over a set of circles as in the proof of Theorem 2.1.1, we are immediately led to the conclusion of this theorem. The second assertion can be proved similarly. *q.e.d.*

One may prove the stronger result that $E(z, s) - y^s \to 0$, or more precisely $E(z, s) - y^s = O(y^{1-s})$, as $y \to \infty$. For this purpose, consider for each cusp on the interval $(0, 1)$ a small circle touching R at the cusp such that the circles do not overlap each other, and note that every point σz, $(\sigma \neq 1)$, is contained in just one such circle when y is large. Similarly, we can prove that $E(z, s) = O(y^{s-1})$ when z tends to a cusp κ which is not equivalent to ∞.

Theorem 2.1.1 and Theorem 2.1.2 can be generalized in various ways. A convenient generalization of them is contained in the following theorem which will be used in Chap. III.

THEOREM 2.1.3. Let $\phi(y)$ be a function of $y > 0$ such that $\phi(y) = O(1)$ as $y \to \infty$ and $\phi(y) = O(y^{1+\varepsilon})$ as $y \to 0$ with some $\varepsilon > 0$. For $z \in H$, put $\phi(z) = \phi(y(z))$. On the other hand, let Γ be a discontinuous group of finite type, let κ be a cusp of Γ, and let $\sigma' \in G$ be such that $\sigma'\infty = \kappa$, $\sigma'^{-1}\Gamma_\kappa\sigma' = \Gamma_0$, where $\Gamma_\kappa = \{\sigma \in \Gamma \mid \sigma\kappa = \kappa\}$ and $\Gamma_0 = \{\begin{pmatrix} 1 & b \\ & 1 \end{pmatrix} \mid b \in \mathbf{Z}\}$. Then, the series $\theta_\phi(z) = \sum \phi(\sigma'^{-1}\sigma z)$, $(\sigma \in \Gamma_\kappa \backslash \Gamma)$, converges absolutely, and is bounded on H.

Proof. Since we have an estimation of the type $|\phi(z)| \leqq c_0 y^{1+\varepsilon}$ with some constant c_0 for all sufficiently small y, every assertion in the theorem is verified by the same arguments as in Theorem 2.1.1 and in Theorem 2.1.2. *q.e.d.*

§2.2. Fourier expansion of an automorphic function

Let Γ be a discontinuous group. A function $f(z)$ is then called an *automorphic function* with respect to Γ, if $f(\sigma z) = f(z)$ holds for all $\sigma \in \Gamma$. If κ is a cusp of Γ, then there exists a $\sigma' \in G$ such that $\sigma'\infty = \kappa$ and such that $\sigma'^{-1}\Gamma_\kappa\sigma' = \Gamma_0$, where $\Gamma_\kappa = \{\sigma \in \Gamma \mid \sigma\kappa = \kappa\}$ is the stabilizer in Γ of κ. If $f(z)$ is an automorphic function, then $f(\sigma' z)$ is a periodic function with period 1, i.e., $f(\sigma'(z + 1)) = f(\sigma' z)$. Therefore, putting

$$e(x) = \exp(2\pi\sqrt{-1}x), \tag{2.2.1}$$

and

$$a_m(y) = \int_0^1 f(\sigma' z)e(-mx)dx, \qquad (z = x + \sqrt{-1}y),$$

we have a Fourier series expansion

$$f(\sigma' z) = \sum_m a_m(y)e(mx), \qquad (m \in \mathbf{Z}),$$

whenever f satisfies some natural analytic conditions. This will be called the *Fourier expansion* of f at the cusp κ. This notion is very important, because our whole theory is based upon a precise investigation of the Fourier expansion of Eisenstein series at a cusp.

Let us now find an explicit formula for the Fourier expansion of an Eisenstein series at a cusp. We denote by Γ a discontinuous group of finite type, and by $\kappa_1, \kappa_2, \ldots, \kappa_h$ a complete set of inequivalent cusps of Γ. The Fourier expansion of $E_i(z, s)$ at κ_j is given in the form

$$E_i(\sigma_j z, s) = \sum_m a_{ij,m}(y, s)e(mx), \qquad (m \in \mathbf{Z}),$$

with

$$a_{ij,m}(y, s) = \int_0^1 E_i(\sigma_j z, s)e(-mx)dx. \tag{2.2.2}$$

To give $a_{ij,m}$ an explicit form, we assume first that Γ is reduced at ∞, and will calculate $a_{11,m}$. However, for the sake of convenience, we omit the index "1" in the notation relative to $\kappa_1 = \infty$; for example we write E for E_1, and a for a_{11}. With this convention, we have firstly

$$a_m(y, s) = \int_0^1 \sum_\sigma y(\sigma z)^s e(-mx)dx, \qquad (\sigma \in \Gamma_0 \backslash \Gamma). \tag{2.2.3}$$

Next, let us make use of the double coset decomposition $\Gamma_0 \backslash \Gamma / \Gamma_0$. Since Γ, as a discontinuous group reduced at ∞, cannot contain any hyperbolic transformation of the form $\begin{pmatrix} t & b \\ & t^{-1} \end{pmatrix}$, $(t > 1)$, we see that the double coset $\Gamma_0 \sigma \Gamma_0$ with $\sigma = \begin{pmatrix} a & b \\ c & d \end{pmatrix} \in \Gamma$ is equal to $\Gamma_0 \begin{pmatrix} 1 & \\ & 1 \end{pmatrix} \Gamma_0$ whenever $c = 0$. Furthermore, we see that, if $c \neq 0$, or more specially if $c > 0$, which may be assumed without any loss of generality, then $\Gamma_0 \sigma \Gamma_0 = \Gamma_0 \sigma' \Gamma_0$ holds for $\sigma = \begin{pmatrix} a & b \\ c & d \end{pmatrix}$, $\sigma' = \begin{pmatrix} a' & b' \\ c' & d' \end{pmatrix}$, if and only if $c = c'$ and $d \equiv d' \pmod{c}$. For a given $c > 0$, there exist only a finite number of d incongruent mod c such that (c, d) is the second row of some $\sigma \in \Gamma$; this follows from Theorem 2.1.1. Thus, we get a decomposition

$$\Gamma_0\backslash\Gamma/\Gamma_0 = \Gamma_0 \cup (\bigcup_{c,d} \Gamma_0\sigma\Gamma_0), \qquad (c > 0,\ d \bmod c,\ \sigma = \begin{pmatrix} * & * \\ c & d \end{pmatrix} \in \Gamma).$$

This is a kind of *Bruhat decomposition.* It also follows from elementary arguments that $\Gamma_0\sigma\sigma_0$ and $\Gamma_0\sigma\sigma_0'$, $(\sigma \notin \Gamma_0)$, are different cosets in $\Gamma_0\backslash\Gamma$, whenever σ_0 and σ_0' are different elements of Γ_0. Hence, $\Gamma_0 \cap \sigma\Gamma_0\sigma^{-1} = \{1\}$ for $c \neq 0$. The right hand side of (2.2.3) is thus equal to

$$\delta_{0m} y^s + \int_{-\infty}^{\infty} \sum_{\sigma} y(\sigma z)^s e(-mx)dx, \qquad (\sigma = \begin{pmatrix} a & b \\ c & d \end{pmatrix} \in \Gamma_0\backslash\Gamma/\Gamma_0,\ c \neq 0),$$

with Kronecker's δ, and the latter is further equal to

$$\delta_{0m} y^s + \sum_{c,d} \int_{-\infty}^{\infty} \frac{y^s}{|cz+d|^{2s}} e(-mx)dx$$

$$= \delta_{0m} y^s + \sum_{c} \frac{1}{|c|^{2s}} \left(\sum_{d} e\left(\frac{md}{c}\right)\right) \cdot \int_{-\infty}^{\infty} \frac{y^s}{|x^2+y^2|^s} e(-mx)dx$$

$$= \delta_{0m} y^s + y^{1-s} \sum \frac{1}{|c|^{2s}} \left(\sum_{d} e\left(\frac{md}{c}\right)\right) \cdot \int_{-\infty}^{\infty} \frac{e(-myt)}{(1+t^2)^s} dt,$$

$$(c > 0,\ d \bmod c,\ \begin{pmatrix} * & * \\ c & d \end{pmatrix} \in \Gamma).$$

We now put

$$(2.2.4) \qquad \sum_{c} \frac{1}{|c|^{2s}} \left(\sum_{d} e\left(\frac{md}{c}\right)\right) = \varphi_m(s), \qquad (c > 0,\ d \bmod c,\ \begin{pmatrix} * & * \\ c & d \end{pmatrix} \in \Gamma),$$

and recall

$$\int_{-\infty}^{\infty} \frac{e(-ut)}{(1+t^2)^s} dt = 2\pi^s |u|^{s-\frac{1}{2}} \Gamma(s)^{-1} K_{s-\frac{1}{2}}(2\pi|u|), \qquad (u \neq 0,\ u \in \boldsymbol{R}),$$

and

$$\int_{-\infty}^{\infty} \frac{dt}{(1+t^2)^s} = \pi^{\frac{1}{2}} \frac{\Gamma(s-\frac{1}{2})}{\Gamma(s)}.$$

Here, K_s means the so-called modified Bessel function defined by

$$K_s(z) = \frac{\pi}{2} \frac{I_{-s}(z) - I_s(z)}{\sin s\pi}$$

with

$$I_s(z) = \sum_{m=0}^{\infty} \frac{(\frac{1}{2}z)^{s+2m}}{m!\,\Gamma(s+m+1)}.$$

(The last series with the additional factor $(-1)^m$ supplied in front of the

mth term for every m, is the series defining of J_s.) We have thus obtained

$$a_m(y, s) = 2\pi^s|m|^{s-\frac{1}{2}}\Gamma(s)^{-1}y^{\frac{1}{2}}K_{s-\frac{1}{2}}(2\pi|m|y)\varphi_m(s), \qquad (m \neq 0), \tag{2.2.5}$$

and

$$a_0(y, s) = y^s + \varphi(s)y^{1-s} \tag{2.2.6}$$

with

$$\varphi(s) = \pi^{\frac{1}{2}}\frac{\Gamma(s-\frac{1}{2})}{\Gamma(s)}\varphi_0(s). \tag{2.2.7}$$

The general case occurring in (2.2.2) can be treated in almost the same way using the double coset decomposition $\Gamma_0\backslash\sigma_i^{-1}\Gamma\sigma_j/\Gamma_0$. The only thing which we need to note is that the coset $\Gamma_0\begin{pmatrix}1 & \\ & 1\end{pmatrix}\Gamma_0$ does not appear in the above decomposition unless $i=j$. For, otherwise, κ_i and κ_j $(i\neq j)$ would be equivalent. The result of the calculations in the general case is

$$a_{ij,m}(y, s) = 2\pi^s|m|^{s-\frac{1}{2}}\Gamma(s)^{-1}y^{\frac{1}{2}}K_{s-\frac{1}{2}}(2\pi|m|y)\varphi_{ij,m}(s), \qquad (m \neq 0), \tag{2.2.8}$$

$$a_{ij,0}(y, s) = \delta_{ij}y^s + \varphi_{ij}(s)y^{1-s}, \tag{2.2.9}$$

with

$$\varphi_{ij,m}(s) = \sum_c \frac{1}{|c|^{2s}}\left(\sum_d e\left(\frac{md}{c}\right)\right), \tag{2.2.10}$$

$$\left(c > 0,\ d \bmod c,\ \begin{pmatrix} * & * \\ c & d\end{pmatrix} \in \sigma_i^{-1}\Gamma\sigma_j\right),$$

and

$$\varphi_{ij}(s) = \pi^{\frac{1}{2}}\frac{\Gamma(s-\frac{1}{2})}{\Gamma(s)}\varphi_{ij,0}(s). \tag{2.2.11}$$

The functions $\varphi_{ij,m}(s)$ are all Dirichlet series in a rather general sense.
The matrix $\Phi(s) = (\varphi_{ij}(s))$ of functions appearing in the constant term of the Fourier expansions of Eisenstein series has a very important meaning, and will be called the *constant term matrix* in the sequel. Here we prove a theorem on $\Phi(s)$.

THEOREM 2.2.1. The constant term matrix $\Phi(s) = (\Phi_{ij}(s))$ in (2.2.11) is a symmetric matrix.

Proof. The sum $\sum_d$ in (2.2.10) is equal to the number of representatives $\sigma = \begin{pmatrix} a & b \\ c & d\end{pmatrix}$ of distinct double cosets in $\Gamma_0\backslash\sigma_i^{-1}\Gamma\sigma_j/\Gamma_0$ with a fixed $c > 0$.

Applying the mapping $G \ni g \to g^{-1} \in G$, we get the same number of representatives of double cosets in $\Gamma_0 \backslash \sigma_j^{-1}\Gamma\sigma_i/\Gamma_0$ with the same c. This implies that $\varphi_{ij}(s) = \varphi_{ji}(s)$. *q.e.d.*

From the fact that E satisfies the differential equation $DE = \lambda E$, it follows from a simple computation that $a_{ij,m}$ in (2.2.2) satisfies

$$\frac{d^2 a_{ij,m}}{dy^2} - \left(4\pi^2 m^2 + \frac{\lambda}{y^2}\right) a_{ij,m} = 0, \qquad (\lambda = s(s-1)). \tag{2.2.12}$$

This again shows that $a_{ij,m}$ is a Bessel function of y.

§2.3. Inner product formulas

If f_1 and f_2 are two automorphic functions with respect to a discontinuous group Γ, we can define the *inner product* (f_1, f_2) by

$$(f_1, f_2) = \int_{\Gamma \backslash H} f_1(z)\overline{f_2(z)}dz \tag{2.3.1}$$

insofar as it exists. It is very interesting that almost all important consequences in our theory come from an investigation of the inner product. For automorphic functions such as Eisenstein series, the inner product can be expressed, to some extent, by means of quantities related to the constant terms of their Fourier expansions. Before stating a fairly general theorem in this direction, we must formulate Green's theorem for the upper half plane H.

Let S be a domain in the complex plane and let ∂S be the boundary of S which is assumed to be a finite union of smooth curves. Then the ordinary form of Green's theorem says

$$\iint_S (f\Delta g - g\Delta f)dxdy = \int_{\partial S}\left(f\frac{\partial g}{\partial n} - g\frac{\partial f}{\partial n}\right)dl$$

for two functions g, f, where $\Delta = \dfrac{\partial^2}{\partial x^2} + \dfrac{\partial^2}{\partial y^2}$, $\dfrac{\partial}{\partial n}$ is the outer normal derivative, and dl is euclidean arc length. From this, we can deduce

$$\iint_S (fDg - gDf)dz = \int_{\partial S}\left(f \cdot y\frac{\partial g}{\partial n} - g \cdot y\frac{\partial f}{\partial n}\right)\frac{dl}{y} \tag{2.3.2}$$

for a subdomain S of H. Here, dl and $\dfrac{\partial}{\partial n}$ are not invariant, while $y\dfrac{\partial}{\partial n}$ resp. dl/y becomes the invariant outer normal derivative on ∂S (resp. the invariant length on H).

As at the beginning of this §, denote now by κ a cusp of a discontinuous group Γ, by Γ_κ the stabilizer in Γ of κ, and by σ' an element of G such that $\sigma'\infty = \kappa$, $\sigma'^{-1}\Gamma_\kappa\sigma' = \Gamma_0$. Furthermore, let $f(z)$ be an automorphic function with respect to Γ which has the Fourier expansion

$$f(\sigma' z) = \sum_m a_m(y)e(mx), \qquad (m \in \boldsymbol{Z}).$$

For a positive number Y, we then define the function $f^Y(z)$ by

$$f^Y(z) = \begin{cases} f(z) - a_0(y(\sigma'^{-1}z)), & \text{if there exists a cusp } \kappa \text{ such that } y(\sigma'^{-1}z) > Y, \\ f(z), & \text{otherwise.} \end{cases}$$

This definition surely makes sense whenever Y is sufficiently large, and then $f^Y(z)$ is again an automorphic function. Roughly speaking, $f^Y(z)$ is obtained by cutting off the constant term of the Fourier expansion at every cusp. We call f^Y the *compact part* of f.

THEOREM 2.3.1. Let Γ be a discontinuous group of finite type, and let $\kappa_1, \kappa_2, \ldots, \kappa_h$ be a complete set of inequivalent cusps of Γ. Assume $f(z, s)$, $g(z, s')$ be two automorphic functions with respect to Γ such that

$$Df(z, s) = \lambda f(z, s), \qquad Dg(z, s') = \lambda' g(z, s'),$$
$$(\lambda = s(s-1), \qquad \lambda' = s'(s'-1),$$

and such that the constant term of the Fourier expansion at κ_i of f, resp. of g, is given by

$$c_i y^s + c_i' y^{1-s}, \quad \text{resp.} \quad d_i y^{s'} + d_i' y^{1-s'},$$

where s and s' are complex constants. Then, for the compact parts f^Y and g^Y of f and g, we have the inner product formula

$$(f^Y(z, s), \overline{g^Y(z, s')})$$
$$= \sum_{i=1}^{h} \left(\frac{c_i d_i Y^{s+s'-1} - c_i' d_i' Y^{-s-s'+1}}{s + s' - 1} + \frac{c_i d_i' Y^{s-s'} - c_i' d_i Y^{-s+s'}}{s - s'} \right)$$

whenever $\lambda \neq \lambda'$ and f^Y and g^Y are square integrable on $\Gamma\backslash H$.

Proof. Let σ_i be as in the definition (2.1.1) of the Eisenstein series, and let

$$f(\sigma_i z) = \sum_{m \in \boldsymbol{Z}} a_{im}(y)e(mx) \quad \text{and} \quad g(\sigma_i z) = \sum_{m \in \boldsymbol{Z}} b_{im}(y)e(mx)$$

be the Fourier expansions of f and g at κ_i. On the other hand, fix a fundamental domain $\mathscr{D} = \Gamma\backslash H$, say, a normal polygon for the sake of simplicity, and assume that $\kappa_1, \kappa_2, \ldots, \kappa_h$ are on the boundary of $\mathscr{D}$. We then denote by S the subdomain of $\mathscr{D}$ consisting of all points $z \in \mathscr{D}$ such that $y(\sigma_i^{-1}z) \leqq Y$ for every i. The segment given as the intersection of $\mathscr{D}$ and the line $y(\sigma_i^{-1}z) = Y$ will be denoted by B_i. So to speak, S is a domain obtained from $\mathscr{D}$ by cutting off neighbourhoods of all cusps (Fig. 4). We now compute the integral

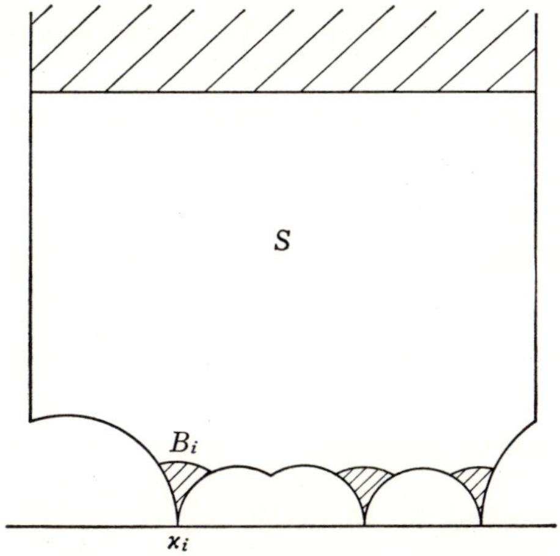

Fig. 4

(2.3.3) $$\iint_S (fDg - gDf)dz = \int_{\partial S}\left(f \cdot y\frac{\partial g}{\partial n} - g \cdot y\frac{\partial f}{\partial n}\right)\frac{dl}{y}$$

which appears in Green's theorem (2.3.2). This integral is equal to

$$\sum_i \int_{Bi}\left(f \cdot y\frac{\partial g}{\partial n} - g \cdot y\frac{\partial f}{\partial n}\right)\frac{dl}{y},$$

because, as we have noted in § 2.1, the set of sides of a normal polygon is divided into the sets of equivalent pairs, and the integral of the outer normal derivative along one side in a pair is equal to the integral of the inner normal derivative along the other side in the pair. To compute $\int_{Bi}$, we omit the index i, and write simply a_m, b_m for a_{im}, b_{im}. Since

$$\frac{d}{dy}(a_m b_{-m}' - a_m' b_{-m}) = a_m b_{-m}'' - a_m'' b_{-m} = -\frac{1}{y^2}(\lambda - \lambda')\, a_m b_{-m}$$

by (2.2.12), we have

$$\int_B \left(f \cdot y\frac{\partial g}{\partial n} - g \cdot y\frac{\partial f}{\partial n}\right)\frac{dl}{y}$$

$$= \int_0^1 \left\{\left(\sum_m a_m(Y)e(mx)\right)\left(\sum_m b_m'(Y)e(mx)\right)\right.$$

$$\left. - \left(\sum_m a_m'(Y)e(mx)\right)\left(\sum_m b_m(Y)e(mx)\right)\right\} dx$$

$$= a_0(Y)b_0'(Y) - a_0'(Y)b_0(Y) + \sum_{m\neq 0}\int_Y^\infty (\lambda - \lambda')\, a_m b_{-m}\frac{dy}{y^2}.$$

After computing explicitly the first term in the last formula, we obtain

$$-(\lambda - \lambda')\,(f^Y, \overline{g^Y})$$

$$= -\sum_i (s - s')\,(c_i d_i Y^{s+s'-1} - c_i' d_i' Y^{-s-s'+1})$$

$$-\sum_i (s + s' - 1)\,(c_i d_i' Y^{s-s'} - c_i' d_i Y^{-s+s'}).$$

This proves the theorem, because $\lambda - \lambda' = (s - s')(s + s' - 1)$. *q.e.d.*

This theorem is called the *Maass-Selberg relation.*

THEOREM 2.3.2. Let Γ be a discontinuous group of finite type, and let $E_i(z, s)$, ($\operatorname{Re} s > 1$, $i = 1, 2, \ldots, h$), be Eisenstein series for a complete set of inequivalent cusps of Γ. Denote by E_i^Y the compact part of E_i, and let $\varphi_{ij}(s)$ be as in (2.2.11). Then, E_i^Y belongs to the space $L^2(\Gamma\backslash H)$ of square integrable functions on $\Gamma\backslash H$, and we have the inner product formula

$$(E_i^Y(z, s), E_j^Y(z, \overline{s'})) = \frac{\delta_{ij} Y^{s+s'-1} - \sum_{l=1}^{h} \varphi_{il}(s)\varphi_{jl}(s')Y^{-s-s'+1}}{s + s' - 1}$$

$$+ \varphi_{ji}(s')\frac{Y^{s-s'}}{s - s'} - \varphi_{ij}(s)\frac{Y^{-s+s'}}{s - s'}, \qquad (\operatorname{Re} s > 1, \quad \operatorname{Re} s' > 1).$$

Proof. The fact that $E_i^Y(z, s) \in L^2(\Gamma\backslash H)$ for $\operatorname{Re} s > 1$ follows from Theorem 2.1.2 and from the fact that $\Gamma\backslash H$ has finite invariant measure. The inner product formula for $s \neq s'$ follows immediately from Theorem 2.3.1. If $s = s'$, then $(E_i^Y(z, s), E_j^Y(z, \bar{s}))$ is equal to the limit of $(E_i^Y(z, s), E_j^Y(z, s'))$, $(s \neq \overline{s'})$, as $s' \to s$, because the proof of the Theorem 2.1.2 shows that the compact part of an Eisenstein series is uniformly bounded in a fundamental domain, if the variable s is restricted to a compact set in the half plane $\operatorname{Re} s > 1$. *q.e.d.*

THEOREM 2.3.3. Let Γ be a discontinuous group of finite type reduced at ∞, and let $E(z, s)$ be the Eisenstein series for ∞. Fix a fundamental domain $\mathscr{D}$ of Γ such that $\mathscr{D}$ is containd in the strip $0 < \operatorname{Re} z < 1$ and such that $\mathscr{D}$ contains the subdomain $\mathscr{D}_Y'$ of the strip determined by $\operatorname{Im} z > Y$ for a sufficiently large $Y > 0$. Define $\tilde{E}^Y(z, s)$ by

$$\tilde{E}^Y(z, s) = \begin{cases} E(z, s) - y^s, & z \in \mathscr{D}_Y', \\ E(z, s), & z \in \mathscr{D}_Y, \end{cases}$$

where $\mathscr{D}_Y = \mathscr{D} - \mathscr{D}_Y'$. Then, we have the inner product formula

$$(\tilde{E}^Y(z, s), \tilde{E}^Y(z, \overline{s'})) = \frac{Y^{s+s'-1}}{s + s' - 1} + \varphi(s')\frac{Y^{s-s'}}{s - s'} - \varphi(s)\frac{Y^{-s+s'}}{s - s'},$$

$$(\operatorname{Re} s > 1, \qquad \operatorname{Re} s' > 1),$$

where $\varphi(s)$ is as in (2.2.7).

Proof. The square integrability of $\tilde{E}^Y$ follows from Theorem 2.1.2. By Theorem 2.3.2, we have

$$(E^Y(z, s), \quad E^Y(z, \overline{s'}))$$

$$= \frac{Y^{s+s'-1} - \sum \varphi_l(s)\varphi_l(s')Y^{-s-s'+1}}{s + s' - 1} + \varphi(s')\frac{Y^{s-s'}}{s - s'} - \varphi(s)\frac{Y^{-s+s'}}{s - s'}$$

where φ_l means the expression φ_{1l} defined previously with respect to a complete set $\kappa_1 = \infty, \kappa_2, \ldots, \kappa_h$ of inequivalent cusps of Γ. The Fourier expansion of $\tilde{E}^Y(z, s)$ at κ_l is $y^{1-s}\varphi_l(s) + \ldots$ for $y > Y$, while the constant term of the Fourier expansion of E^Y at κ_l is 0 for $y > Y$. On the other hand, we have

$$\int_Y^\infty y^{1-s}\varphi_l(s) \cdot y^{1-s'}\varphi_l(s')\frac{dy}{y^2} = \varphi_l(s)\varphi_l(s')\frac{Y^{-s-s'+1}}{s + s' - 1}$$

Adding this term to $(E^Y(z, s), E^Y(z, \bar{s}'))$ for all $l = 1, 2, \ldots, h$, one gets the conclusion of the theorem. *q.e.d.*

CHAPTER III

Analytic continuation of Eisenstein series (First step)

§3.1. Continuation of the constant term

In previous chapters, the complex variable s of an Eisenstein series $E(z, s)$ was always restricted by the condition $\mathrm{Re}\, s > 1$. In this chapter, however, we shall obtain the analytic continuation of $E(z, s)$ to the domain $\mathrm{Re}\, s > \frac{1}{2}$ minus the interval $(\frac{1}{2}, 1]$. For this purpose, one has first to obtain the continuation of the functions $\varphi_{ij}(s)$ appearing in the constant term matrix of the Eisenstein series.

Let $n(x) = \begin{pmatrix} 1 & x \\ & 1 \end{pmatrix}$ be as in (1.1.1), and let $\psi(z)$ be a function on the upper half plane H such that $\psi(n(x)z) = \psi(z)$ for all $x \in R$. We define then the *Mellin transform* $L_\psi(z, s)$ of ψ by

(3.1.1)
$$L_\psi(z, s) = \int_0^\infty \psi(a(y)z) y^{-s} \frac{dy}{y},$$

where $a(y) = \begin{pmatrix} \sqrt{y} & \\ & \sqrt{y}^{-1} \end{pmatrix}$, $(y > 0)$, is again as in (1.1.1). The inverse transformation of (3.1.1) is given by the complex line integral

(3.1.2)
$$\psi(a(y)z) = \frac{1}{2\pi\sqrt{-1}} \int_{\mathrm{Re}\, s = S_0} L_\psi(z, s) y^s ds,$$

where S_0 is to be chosen suitably according to the nature of the function $L_\psi(z, s)$ of s. However, in the particular case when $\psi(a(y)z)$ decreases more rapidly than any positive power of y^{-1} (resp. y) when $y \to \infty$ (resp. $y \to 0$), $L_\psi(z, s)$ is an entire function of s, and S_0 in (3.1.2) may be taken arbitrarily. The value $L_\psi(\sqrt{-1}, s)$ will be denoted simply by $L_\psi(s)$, and, if no confusion is possible, $\psi(a(y)\sqrt{-1})$ will be denoted by $\psi(y)$, so that $\psi(z) = \psi(y(z))$ with $y(z) = \mathrm{Im}\, z$, and one has $L_\psi(z, s) = L_\psi(s) y(z)^s$ and

$$L_\psi(s) = \int_0^\infty \psi(y) y^{-s} \frac{dy}{y}.$$

Let Γ be a discontinuous group of finite type and let $\kappa_1, \kappa_2, \ldots, \kappa_h$ be a complete set of inequivalent cusps. For a function $\psi(z)$ as above, we put

$$\theta_{i,\psi}(z) = \sum_\sigma \psi(\sigma_i^{-1}\sigma z), \qquad (\sigma \in \Gamma_i \backslash \Gamma), \tag{3.1.3}$$

where the meanings of σ_i, Γ_i are as in the definition (2.1.1) of the Eisenstein series. If Γ is reduced at ∞, and if $\kappa_1 = \infty$, then we put simply

$$\theta_\psi(z) = \sum_\sigma \psi(\sigma z), \qquad (\sigma \in \Gamma_0 \backslash \Gamma),$$

with $\Gamma_0 = \{n(b) \mid b \in \boldsymbol{Z}\}$.

We now fix a discontinuous group Γ reduced at ∞, and will show how one can obtain an analytic continuation to the domain $\operatorname{Re} s > 1/2$, except for the interval $(1/2, 1]$, of the function $\varphi(s)$ appearing in the Fourier expansion of the Eisenstein series $E(z, s)$ for the cusp ∞.

THEOREM 3.1.1. Let ψ, ψ' be two functions on H satisfying $\psi(n(x)z) = \psi(z)$, $\psi'(n(x)z) = \psi'(z)$ for all $n(x) = \begin{pmatrix} 1 & x \\ & 1 \end{pmatrix}$, $(x \in \boldsymbol{R})$, and let Γ be a discontinuous group of finite type reduced at ∞. Assume that $\psi(z)$, $(z = x + \sqrt{-1}y)$, decreases sufficiently rapidly both when $y \to \infty$ and when $y \to 0$ that the integral in (3.1.1) be absolutely convergent for all s, and assume that ψ' has the same property for $\operatorname{Re} s = S_0$ for some $S_0 > 1$. Then, θ_ψ belongs to the space $L^2(\Gamma \backslash H)$ of square integrable functions on $\mathscr{D} = \Gamma \backslash H$, and the inner product formula

$$(\theta_\psi, \bar{\theta}_{\psi'}) = \frac{1}{2\pi\sqrt{-1}} \int_{\operatorname{Re} s = S_0} L_\psi(1 - s) L_{\psi'}(s)\, ds$$
$$+ \frac{1}{2\pi\sqrt{-1}} \int_{\operatorname{Re} s = S_0} L_\psi(s) L_{\psi'}(s) \varphi(s)\, ds$$

holds provided that $\theta_{\psi'} \in L^2(\Gamma \backslash H)$, where $\varphi(s)$ is as in (2.2.7).

Proof. In a manner similar to that of the proof of Theorem 2.1.2, it may be proved that $\theta_\psi(z)$ is bounded on H. Therefore, $\theta_\psi \in L^2(\Gamma \backslash H)$.

Let now $E(z, s)$ be the Eisenstein series for the cusp ∞ of Γ. Then,

$$(\theta_\psi, \bar{\theta}_{\psi'}) = \int_{\mathscr{D}} \theta_\psi(z) \theta_{\psi'}(z) dz = \int_0^\infty \int_0^1 \psi(z) \theta_{\psi'}(z) \frac{dxdy}{y^2},$$

and, since

$$\psi'(z) = \frac{1}{2\pi\sqrt{-1}}\int_{\mathrm{Re}\,s=S_0} L_{\psi'}(s)y^s ds, \qquad (z = x + \sqrt{-1}y),$$

by (3.1.2), we have

$$\theta_{\psi'}(z) = \frac{1}{2\pi\sqrt{-1}}\int_{\mathrm{Re}\,s=S_0} L_{\psi'}(s)E(z, s)ds$$

Hence, using the Fourier expansion of $E(z, s)$ and (2.2.6), we have

$$(\theta_\psi, \overline{\theta}_{\psi'}) = \frac{1}{2\pi\sqrt{-1}}\int_{\mathrm{Re}\,s=S_0} L_{\psi'}(s)\int_0^\infty \psi(y)(y^s + \varphi(s)y^{1-s})\frac{dy}{y^2}\,ds$$

$$= \frac{1}{2\pi\sqrt{-1}}\int_{\mathrm{Re}\,s=S_0} (L_\psi(1-s)L_{\psi'}(s) + L_\psi(s)L_{\psi'}(s)\varphi(s))ds,$$

as stated in the theorem. *q.e.d.*

It is now important to recall some facts about the Fourier transformation. We define the *Fourier transform* $\hat{f}$ of a function $f(x)$ defined on $\boldsymbol{R}$ by putting

$$\hat{f}(y) = \int_{-\infty}^\infty f(x)e(xy)dx, \qquad (e(x) = \exp(2\pi\sqrt{-1}x)).$$

The inverse transformation is given by

$$f(x) = \int_{-\infty}^\infty \hat{f}(y)e(-xy)dy. \tag{3.1.4}$$

The Mellin transformation in (3.1.1) is essentially the same thing as the Fourier transformation, because

$$L_\psi(S + \sqrt{-1}t) = \int_0^\infty \psi(y)y^{-s}\frac{dy}{y} = \int_{-\infty}^\infty \psi(e^{y'})e^{-Sy'}e^{-\sqrt{-1}ty'}dy'$$

for $s = S + \sqrt{-1}t$. The formula (3.1.2) for the inverse Mellin transformation is an immediate consequence of (3.1.4).

THEOREM 3.1.2. Let $h(s)$ be an analytic function of a complex variable s, and suppose that the Fourier transform of $h(S + \sqrt{-1}t)$ with respect to $t \in \boldsymbol{R}$ is a bounded function for any fixed S such that $S_0 - \varepsilon < S < S_1 + \varepsilon$, where S_0, S_1 and $\varepsilon > 0$ are real constants. Then,

$$\psi(y) = \frac{1}{2\pi\sqrt{-1}}\int_{\mathrm{Re}\,s=S} h(s)y^s ds$$

has the property $\psi(y) = O(y^{S_1})$ as $y \to 0$, and $\psi(y) = O(y^{S_0})$ as $y \to \infty$.

Proof. Since $h(s)$ is holomorphic in the region $S_0 - \varepsilon < S < S_1 + \varepsilon$, the function $\psi(y)$ does not depend on S.

Now, we have

$$\frac{1}{2\pi\sqrt{-1}}\int_{\mathrm{Re}\, s=S_1} h(s)y^s ds = \frac{1}{2\pi}\int_{-\infty}^{\infty} h(S_1+\sqrt{-1}t)y^{S_1+\sqrt{-1}t}dt$$

$$= y^{S_1}\cdot\frac{1}{2\pi}\int_{-\infty}^{\infty} h(S_1+\sqrt{-1}t)e\Big(\frac{t\log y}{2\pi}\Big)dt.$$

This proves the first assertion. The second assertion is proved simply by replacing S_1 with S_0. *q.e.d.*

A function f on $\boldsymbol{R}$ is said to belong to the *Schwartz space* $\mathscr{S}(\boldsymbol{R})$ on $\boldsymbol{R}$, if $x^m\dfrac{d^n}{dx^n}f(x)$ is bounded on $\boldsymbol{R}$ for every non-negative $m, n \in \boldsymbol{Z}$. A basic property of $\mathscr{S}(\boldsymbol{R})$ is that the Fourier transformation maps $\mathscr{S}(\boldsymbol{R})$ onto itself. To get the analytic continuation of the constant term matrix $\Phi(s)$ of the Eisenstein series, we shall make use of an entire function $h(s)$ such that the function $h(S+\sqrt{-1}t)$ of $t \in \boldsymbol{R}$ belongs to $\mathscr{S}(\boldsymbol{R})$ for any $S \in \boldsymbol{R}$. An example of such a function is e^{s^2}.

Before treating the analytic continuation of $\Phi(s)$, we need one more preparatory result on Hilbert space. As is verified easily by partial differentiation, the Laplacian $D = y^2\Big(\dfrac{\partial^2}{\partial x^2}+\dfrac{\partial^2}{\partial y^2}\Big)$ is a self-adjoint operator on the Hilbert space $\mathfrak{H} = L^2(\Gamma\backslash H)$, i.e., we have $(Df_1, f_2) = (f_1, Df_2)$ for any two automorphic functions for which D is defined; thus, D is a closed operator on $\mathfrak{H}$. Therefore, if $\lambda \in \boldsymbol{C}$, then $D + \mathrm{Re}\lambda$ is again self-adjoint, and we have

$$\begin{aligned}||(D+\lambda)f||^2 &= ||(D+\mathrm{Re}\,\lambda)f + \sqrt{-1}\,\mathrm{Im}\,\lambda\cdot f||^2\\ &= ||(D+\mathrm{Re}\,\lambda)f||^2 + ||\mathrm{Im}\,\lambda\cdot f||^2,\end{aligned}$$

so that

$$||(D+\lambda)f|| \geqq |\mathrm{Im}\,\lambda|\;||f||. \tag{3.1.5}$$

It is known from the general theory of Hilbert space that (3.1.5) with $\mathrm{Im}\,\lambda \neq 0$ implies the existence of a bounded operator R_λ on $\mathfrak{H}$ such that $R_\lambda\cdot(D+\lambda)$ is the identity. This R_λ is called the *resolvent* of D. The resolvent R_λ is, so to speak, the inverse operator of $D+\lambda$. Using the terminology of Hilbert space further, the existence of R_λ implies that λ does not belong to the *spectrum* of D, and that R_λ gives a *holomorphic family of operators* when λ moves without touching the spectrum of D; thus, $(f_1, \overline{R_\lambda f_2})$ is a holomorphic function of λ.

THEOREM 3.1.3. Let Γ be a discontinuous group of finite type reduced at ∞, let $E(z, s)$ be the Eisenstein series for the cusp ∞ of Γ, and let $y^s + \varphi(s)y^{1-s}$ be the constant term of the Fourier expansion of $E(z, s)$ at ∞. Then, $\varphi(s)$ may be continued holomorphically to the domain determined by $\operatorname{Re} s > 1/2$, $s \notin (1/2, 1]$.

Proof. Let $h(s)$ be an entire function such that the function $h(S + \sqrt{-1}t)$ of $t \in \boldsymbol{R}$ belongs to the Schwartz space $\mathscr{S}(\boldsymbol{R})$ for any fixed $S \in \boldsymbol{R}$. For instance, take $h(s) = e^{s^2}$. Denote by $\psi(y)$ the function of $y > 0$ given by

$$\psi(y) = \frac{1}{2\pi\sqrt{-1}} \int_{\operatorname{Re} s = S} h(s) y^s ds.$$

Then, it follows from Theorem 3.1.2 that $\psi(y)$ decreases more rapidly than any power of y both as $y \to 0$ and as $y \to \infty$, so that $\theta_\psi(z) \in L^2(\Gamma\backslash H)$ by Theorem 2.1.3. Next let ζ be a complex number such that $\operatorname{Re} \zeta > 1$, and define the function $\psi_\zeta(y)$ by

$$\psi_\zeta(y) = \frac{1}{2\pi\sqrt{-1}} \int_{\operatorname{Re} s = S_0} \frac{h(s)}{(s-\zeta)(s-(1-\zeta))} y^s ds,$$

where S_0 is a real number such that $\operatorname{Re}(1-\zeta) < S_0 < \operatorname{Re} \zeta$. Then, Theorem 3.1.2 shows that $\psi_\zeta(y) = O(y^{1+\varepsilon})$ as $y \to 0$, and $\psi_\zeta(y) = O(y^{-\varepsilon})$ as $y \to \infty$ for any $\varepsilon > 0$. Therefore, again from Theorem 2.1.3, it follows that $\theta_{\psi_\zeta}(y) \in L^2(\Gamma\backslash H)$. For the sake of simplicity, let us put here $\theta_\psi = \theta$ and $\theta_{\psi_\zeta} = \theta_\zeta$, and assume $\operatorname{Re} s = S_0 > 1$. Then,

$$\theta_\zeta(z) = \frac{1}{2\pi\sqrt{-1}} \int_{\operatorname{Re} s = S_0} \frac{h(s)}{(s-\zeta)(s-(1-\zeta))} E(z, s) ds,$$

and hence

$$(D - \zeta(\zeta - 1))\theta_\zeta(z) = \theta(z).$$

This means that $\theta_\zeta = R_\lambda \theta$, where R_λ is the inverse operator of $D + \lambda$, $(\lambda = -\zeta(\zeta - 1))$, that is, the resolvent of D.

Now, Theorem 3.1.1 shows that

$$(\theta, \bar{\theta}_\zeta) = \frac{1}{2\pi\sqrt{-1}} \int_{\operatorname{Re} s = S_0} h(1-s) \frac{h(s)}{(s-\zeta)(s-(1-\zeta))} ds$$
$$+ \frac{1}{2\pi\sqrt{-1}} \int_{\operatorname{Re} s = S_0} h(s) \frac{h(s)}{(s-\zeta)(s-(1-\zeta))} \varphi(s) ds.$$

Therefore, S_1 being a real number such that $\operatorname{Re} \zeta < S_1$, we see

$$(\theta, \overline{R_\lambda \theta}) = \frac{1}{2\pi\sqrt{-1}} \int_{\mathrm{Re}\, s = s_1} \frac{h(1-s)h(s) + h(s)^2 \varphi(s)}{(s-\zeta)(s-(1-\zeta))} ds - \frac{h(1-\zeta)h(\zeta)}{2\zeta - 1} - \frac{h(\zeta)^2}{2\zeta - 1} \varphi(\zeta).$$

This formula allows ζ to move beyond the region $\mathrm{Re}\, \zeta > 1$. In fact, the integral and the two fractions on the right hand side are certainly well-defined whenever $\mathrm{Re}\, \zeta > 1/2$. Moreover, R_λ gives a holomorphic family of linear operators of $L^2(\Gamma/H)$, whenever $\zeta(\zeta - 1)$ does not touch the real line. This means in particular that $(\theta, \overline{R_\lambda \theta})$ is a holomorphic function, whenever ζ satisfies the conditions $\mathrm{Re}\, \zeta > 1/2$ and $\mathrm{Im}\, \zeta \neq 0$. *q.e.d.*

We now consider an arbitrary discontinuous group Γ of finite type for which $\kappa_1, \kappa_2, \dots, \kappa_h$ is a complete set of inequivalent cusps, and propose to prove the existence of a holomorphic continuation of the constant term matrix $\Phi(s)$. What we have to do for this purpose is essentially the same thing as in the proof of Theorem 3.1.3. Instead of θ_ψ, we use $\theta_{i,\psi}$, defined as in (3.1.3), and obtain

$$(\theta_{i,\psi}, \overline{\theta_{j,\psi'}}) = \frac{\delta_{ij}}{2\pi\sqrt{-1}} \int_{\mathrm{Re}\, s = s_0} L_\psi(1-s) L_{\psi'}(s) ds + \frac{1}{2\pi\sqrt{-1}} \int_{\mathrm{Re}\, s = s_0} L_\psi(s) L_{\psi'}(s) \varphi_{ij}(s) ds,$$

which is a generalization of the inner product formula in Theorem 3.1.1; and, by specializing ψ to be the same function as in the proof of Theorem 3.1.3, we obtain

$$(3.1.6) \qquad (\theta_{i,\psi}, \overline{R_\lambda \theta_{j,\psi}}) = \frac{1}{2\pi\sqrt{-1}} \int_{\mathrm{Re}\, s = s_1} \frac{\delta_{ij} h(1-s)h(s) + h(s)^2 \varphi_{ij}(s)}{(s-\zeta)(s-(1-\zeta))} ds - \frac{\delta_{ij} h(1-\zeta)h(\zeta)}{2\zeta - 1} - \frac{h(\zeta)^2}{2\zeta - 1} \varphi_{ij}(\zeta)$$

instead of the formula for $(\theta, \overline{R_\lambda \theta})$ given in the proof of Theorem 3.1.3. In this way, the following theorem is proved:

THEOREM 3.1.4. Let Γ be a discontinuous group of finite type. Then, each function $\varphi_{ij}(s)$ in the constant term matrix $\Phi(s)$ of the Eisenstein series for Γ may be continued holomorphically to the domain determined by $\mathrm{Re}\, s > 1/2$, $s \notin (1/2, 1]$.

§3.2. Continuation of the Eisenstein series

Our next aim is to find the analytic continuation of Eisenstein series to the same domain on which the constant term matrix has a holomorphic continuation by Theorem 3.1.3. First of all, it is necessary to state the following important theorem based upon the properties of Bessel functions.

THEOREM 3.2.1. Let $f(z)$ be an automorphic function with respect to a discontinuous group Γ of finite type. Suppose that $Df = \lambda f$, $(\lambda \in \mathbf{C})$, where D is the Laplacian, and that the Fourier expansion of f at any cusp of Γ is of the form

$$c_0 y^s + c_0' y^{1-s} + \sum_{m \neq 0} c_m y^{\frac{1}{2}} K_{s-\frac{1}{2}}(2\pi|m|y)e(mx), \qquad (m \in \mathbf{Z}), \tag{3.2.1}$$

where s is a complex constant with $s(s-1) = \lambda$, c_m are also constants, and the other notation is as in §2.2. Then, the compact part f^Y of f, in the sense of §2.3, is bounded.

Proof. We may assume that Γ is reduced at ∞, and it is enough to examine the behavior of $f(z)$ when z tends to the cusp ∞ along a curve in a fundamental domain $\mathscr{D} = \Gamma\backslash H$ of the form indicated in Fig. 4, §2.3.

The theory of Bessel functions shows that the absolute value of $\alpha(y) = y^{\frac{1}{2}} K_{s-\frac{1}{2}}(2\pi y)$ is a monotonically decreasing function for $y \geqq y_0$, where $y_0 > 0$ is a constant depending on s. One easy way to draw this conclusion is first to apply repeatedly the partial integration which gives

$$\int_{-\infty}^{\infty} \frac{e(-ut)}{(1+t^2)^s}\,dt = -\frac{s}{\pi\sqrt{-1}u}\int_{-\infty}^{\infty} \frac{te(-ut)}{(1+t^2)^{s+1}}\,dt$$

in order to show that $\alpha(y) \to 0$ as $y \to \infty$, and then to use the differential equation

$$\frac{d^2\alpha}{dy^2} - \left(4\pi^2 + \frac{\lambda}{y^2}\right)\alpha = 0,$$

(*cf.*(2.2.12)), to show that Re α and Im α have the same sign as their second order derivatives for large y.

Thus, it follows from the properties of Fourier series that (3.2.1) is absolutely convergent for any $y > 0$, and that $\sum |c_m y^{\frac{1}{2}} K_{s-\frac{1}{2}}(2\pi|m|y)|$ is monotonically decreasing as $y \to \infty$, $(y \geqq y_0)$. *q.e.d.*

This theorem may be generalized easily to the case where s is a variable restricted in a compact set S and c_m depends holomorphically on s. In this case, $b(y, s) = \sum |c_m(s) y^{\frac{1}{2}} K_{s-\frac{1}{2}}(2\pi|m|y)|$ is a real analytic function, and hence continuous with respect to s. Therefore, $|b(y, s)| \leqq |b(y_0, s)| \leqq |b(y_0, s_0)|$ with some $y_0 > 0$ depending on S and some $s_0 \in S$. In particular, we have

COROLLARY. Let S be a compact subset of $\boldsymbol{C}$ on which an Eisenstein series $E(z, s)$ for a cusp of a discontinuous group Γ of finite type has a holomorphic continuation. Then, the compact part $E^Y(z, s)$ of $E(z, s)$ is uniformly bounded on S.

Here we recall some basic facts from the theory of functions of several complex variables. It suffices for our purpose to investigate functions $f(s, s')$ of two complex variables s, s'. For the Taylor series

$$\sum a_{m,m'}(s - s_0)^m(s' - s_0')^{m'}, \qquad (m, m' = 0, 1, 2, \ldots),$$

with

$$a_{m,m'} = \frac{1}{m!m'!}\frac{\partial^{m+m'}}{\partial^m s \partial^{m'} s'} f(s, s')\bigg|_{s=s_0,\ s'=s_0'},$$

there is no radius of convergence as in the case of one complex variable, but the Taylor series is absolutely convergent for all s, s' with $|s - s_0| \leqq |s_1 - s_0|$, $|s' - s_0'| \leqq |s_1' - s_0'|$ whenever the series is convergent for $s = s_1$, $s' = s_1'$. Conversely, if $f(s, s')$ is holomorphic in the domain determined by $|s - s_0| \leqq |s_1 - s_0|$, $|s' - s_0'| \leqq |s_1' - s_0'|$ for some pair of complex numbers s_1, s_1', then the Taylor series is absolutely convergent in the same domain. Again provided that the Taylor series for $f(s, s')$ is convergent at $s = s_1$, $s' = s_1'$ with $|s - s_1| = r$, $|s' - s_1'| = r'$, the term $a_{m,m'}r^m r'^{m'}$ tends to 0 as m or m' tends to ∞. Various relations between r, r' and coefficients $a_{m,m'}$ may be deduced from this fact. For example, putting $m = m'$, we have

$$\text{(3.2.2.)} \qquad \overline{\lim_{m\to\infty}}\left|\frac{1}{(m!)^2}\frac{\partial^{2m}}{\partial^m s \partial^m s'} f(s, s')\right|^{1/2m}_{s=s_0,\ s'=s_0'} \leqq (\min(r, r'))^{-1}.$$

THEOREM 3.2.2. Let Γ be a discontinuous group of finite type, and let $\kappa_1, \kappa_2, \ldots, \kappa_h$ be a complete set of inequivalent cusps of Γ. Then, the Eisenstein series $E_i(z, s)$ for each cusp κ_i has a holomorphic continuation with respect to s to the domain determined by $\operatorname{Re} s > ½$, $s \notin (½, 1]$.

Proof. We may assume that Γ is reduced at ∞, and that $E_i(z, s)$ is the Eisenstein series $E(z, s)$ for the cusp $\kappa_1 = \infty$. The function φ_{11} in the constant term matrix $\Phi(s)$ will be denoted by φ.

Let $s_0 = S_0 + \sqrt{-1}t_0$ be a complex number such that $\operatorname{Re} s_0 = S_0 > 1$, $t_0 > 0$. Then, $E(z, s)$ has a Taylor expansion

$$\text{(3.2.3)} \qquad E(z, s) = \sum_{n=0}^{\infty} \frac{1}{n!} E^{(n)}(z, s_0)(s - s_0)^n$$

with $E^{(n)}(z, s) = \dfrac{d^n}{ds^n} E(z, s)$. We propose to estimate the radius of convergence of this Taylor series.

It follows from Theorem 2.3.3, with $s = S + \sqrt{-1}t$, $s' = S' - \sqrt{-1}t$, that

$$\begin{aligned}&(\tilde{E}^Y(z, S+\sqrt{-1}t),\ \tilde{E}^Y(z, S'-\sqrt{-1}t))\\ &= \frac{Y^{S+S'-1}}{S+S'-1} + \varphi(S'-\sqrt{-1}t)\frac{Y^{S-S'+2\sqrt{-1}t}}{S-S'+2\sqrt{-1}t}\\ &\quad - \varphi(S+\sqrt{-1}t)\frac{Y^{-S+S'-2\sqrt{-1}t}}{S+S'-2\sqrt{-1}t}\end{aligned}$$

For a moment, let us regard the right hand side of this formula as a function $f(S, S')$ of two complex variables. Then, Theorem 3.1.4. shows in particular that $f(S, S')$ is holomorphic whenever $|S - S_0| < \min(S_0 - \tfrac{1}{2}, t_0)$, $|S' - S_0| < \min(S_0 - \tfrac{1}{2}, t_0)$, because the circle $|S - S_0| = \min(S_0 - \tfrac{1}{2}, t_0)$ touches neither the real axis nor the line Re $s = \tfrac{1}{2}$. Hence, (3.2.2) implies

$$\overline{\lim_{m\to\infty}}\left|\frac{1}{(m!)^2}\frac{\partial^{2m}}{\partial S^m \partial S'^m}(\tilde{E}^Y(z, s+\sqrt{-1}t),\ \tilde{E}^Y(z, S'+\sqrt{-1}t))\right|^{1/2m}_{S=S'=S_0,t=t_0}$$
$$\leqq \left(\min\left(S_0 - \frac{1}{2}, t_0\right)\right)^{-1}$$

so that

$$\overline{\lim_{n\to\infty}}\left\|\frac{1}{n!}\tilde{E}_{(n)}{}^Y(z, s_0)\right\|^{1/n} \leqq \left(\min\left(S_0 - \frac{1}{2}, t_0\right)\right)^{-1},$$

where $\tilde{E}_{(n)}{}^Y(z, s) = \dfrac{d^n}{ds^n}\tilde{E}^Y(z, s)$. This implies that the Taylor series

$$\sum_{n=0}^{\infty}\frac{1}{n!}\,||\tilde{E}_{(n)}{}^Y(z, s_0)||\,(s - s_0)^n$$

converges whenever $|s - s_0| < \min(S_0 - \tfrac{1}{2}, t_0)$, and therefore the serie[s]

$$\sum_{n=0}^{\infty}\frac{1}{n!}\tilde{E}_{(n)}{}^Y(z, s_0)(s - s_0)^n \tag{3.2.4}$$

for each of the same values of s is convergent with respect to the $L^2(\Gamma\backslash H)$-norm, i.e., strongly convergent in the Hilbert space $L^2(\Gamma\backslash H)$. Therefore, it follows from Theorem 1.3.4 that the series (3.2.4) with $|s - s_0| < \min(S_0 - \tfrac{1}{2}, t_0)$ converges to a function in $L^2(\Gamma\backslash H)$ which is an eigenfunction of D except when $y = Y$. (To apply Theorem 1.3.4, consider a subdomain of $\mathscr{D} = \Gamma\backslash H$ which does not cross the line $y = Y$, and put $\lambda_1 = \lambda_2 = \ldots = \lambda = s(s-1)$.) This means that (3.2.3) is also convergent pointwise if $|s - s_0| < \min(S_0 - \frac{1}{2}, t_0)$. Of course, the situation is the same for $t_0 < 0$. By virtue of these results we see that $E(z, s)$ is holomorphic whenever s is in the domain consisting of those points satisfying at least one of the following four conditions: 1) $\text{Re}\, s > 1$, 2) $\text{Re}\, s > \frac{1}{2}$, $|\text{Im}\, s| > \frac{1}{2}$, 3) $|s - (1 + \frac{1}{2}\sqrt{-1})| < \frac{1}{2}$, 4) $|s - (1 - \frac{1}{2}\sqrt{-1})| < \frac{1}{2}$ (Fig. 5). Now let s_0, with $\text{Re}\, s_0 < 1$, be in this domain and

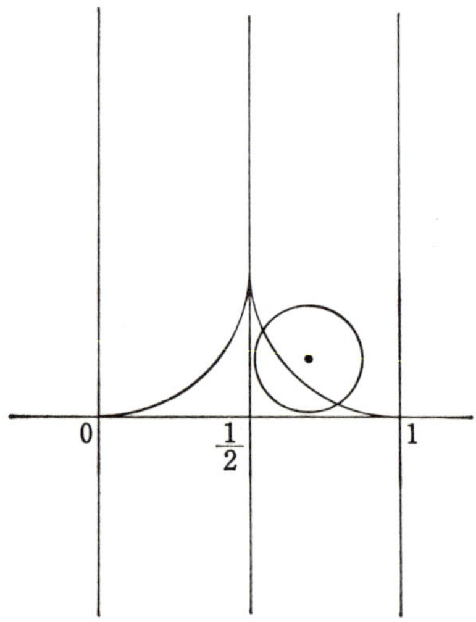

Fig. 5

let r be a positive number such that $r < \min(\text{Re}\, s_0 - \frac{1}{2}, |\text{Im}\, s_0|)$. Then, by the Corollary to Theorem 3.2.1, and from the fact that the norm $||E_{(n)}{}^Y(z, s)||$ is a real analytic function of $\text{Re}\, s$ and of $\text{Im}\, s$, it follows that all the previous steps in this proof can be repeated under the assumption that $|s - s_0| < r$, even if the circle $|s - s_0| = r$ does not lie in the domain determined by the above four conditions. Thus, we obtain a larger domain in the complex plane on which $E(z, s)$ is holomorphic as a function of s. Since the same process of analytical continuation may be repeated as often as desired, one obtains the assertion of the theorem. *q.e.d.*

For the proof of this theorem, it is enough to know that one diagonal element $\varphi_{ii}(s)$ of the constant term matrix $\Phi(s)$ can be continued holomorphically. Therefore, Theorem 3.1.4 in the proof of Theorem 3.2.2 can be replaced by Theorem 3.1.3, and then Theorem 3.1.4 follows from Theorem 3.2.2.

It may be proved in standard fashion that the Eisenstein series obtained above by analytical continuation with respect to s is an automorphic function, because the original Eisenstein series defined on the smaller domain had the same properties.

It is already obtained in the proof of Theorem 3.2.2 that, even after the analytic continuation, the Eisenstein series is an eigenfunction of D with the eigenvalue $\lambda = s(s - 1)$.

CHAPTER IV

Analytic continuation of Eisenstein series (Second step)

§4.1. Preparations

The purpose of this chapter is to obtain an analytic continuation of the Eisenstein series to the region determined by $\operatorname{Re} s \geqq \frac{1}{2}$; this result will immediately furnish the analytic continuation to the whole complex s-plane. The arguments for the analytic continuation to $\operatorname{Re} s \geqq \frac{1}{2}$ are essentially elementary, but in a certain sense highly technical; we must complete each step carefully.

In this §, we state several auxiliary theorems on which the proofs given in the subsequent articles of this chapter are based.

We will always use $f(z, s)$ to denote an automorphic functing of z with respect to a discontinuous group Γ of finite type operating on the upper half plane H, such that $f(z, s)$ depends analytically on the complex parameter s and satisfies

$$Df(z, s) = \lambda f(z, s), \qquad (\lambda = s(s-1)), \tag{4.1.1}$$

with $D = y^2\left(\frac{\partial^2}{\partial x^2} + \frac{\partial^2}{\partial y^2}\right)$ being the Laplacian of H. Moreover, $\kappa_1, \kappa_2, \ldots, \kappa_h$ being a complete set of inequivalent cusps of Γ, we assume that the constant term of the Fourier expansion of $f(z, s)$ at κ_i is of the form

$$c_i(s)y^s + c_i'(s)y^{1-s}, \tag{4.1.2}$$

($c_i'(s)$ is not the derivative of $c_i(s)$).

The definition of the inner product of automorphic functions is to be taken as in (2.3.1), and all automorphic functions in this § are assumed to have compact parts belonging to the Hilbert space $L^2(\Gamma\backslash H)$ of square integrable automorphic functions. In other words, the automorphic functions in this § are of the type mentioned in Theorem 3.2.1.

THEOREM 4.1.1. Let $f(z, s)$ be an automorphic function of z as in (4.1.1), where s belongs to a given set M of complex numbers. Let $\{s_n'\}$ be a sequence of complex numbers in M converging to $s_0 \in \mathbf{C}$, and assume that the norm of the compact part $f^Y(z, s_n')$ for any fixed Y is bounded by a single constant for all n. Then, there exists a subsequence $\{s_n\}$ of $\{s_n'\}$ such that $f(z, s_n)$ converges pointwise to an automorphic function $f(z)$ satisfying $Df = \lambda_0 f$ with $\lambda_0 = s_0(s_0 - 1)$. Furthermore, for any such subsequence, the absolute value of $f^Y(z, s_n)$ is bounded by a single constant for all n and z, and $f^Y(z, s_n)$ converges strongly to $f^Y(z)$.

Proof. Consider first a weakly convergent subsequence of $\{f^Y(z,s_n')\}$. Then, the conclusion about pointwise convergence follows from Theorem 1.3.4.

Next we assume that $f(z, s_n)$ is a sequence, as specified in the theorem, which converges to $f(z)$, and fix a point pair invariant $k(z, z')$ such that $k(z, z') = 0$ or 1 according as the distance $d(z, z')$ between z and z' is greater than a given fixed constant C or not. We have

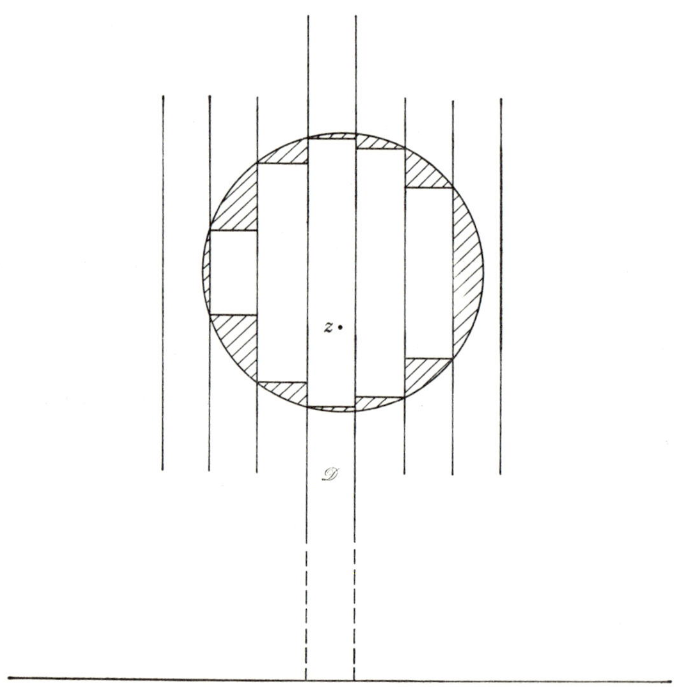

Fig. 6

$$\int_H k(z, z')f(z', s_n)dz' = \Lambda(s_n)f(z, s_n) \tag{4.1.3}$$

with $\Lambda(s_n) \to \Lambda(s_0)$, where $\Lambda(s_0)$ may be assumed to be a non-zero constant. This and Schwarz' inequality imply the boundedness of $f^Y(z, s_n)$ on every compact subset of $\mathscr{D}$. If z is near a cusp of $\mathscr{D}$, then the disc $\{z' \mid d(z, z') < C\}$ lies completely in the region in which the constant term of the Fourier expansion of $f^Y(z,s_n)$ at the cusp is 0. Therefore, we have an integral representation of $f^Y(z,s_n)$, obtained by replacing f with f^Y in (4.1.3). This integral reduces to an integral of $f^Y(z,s_n)$ on a subset $S(\varepsilon)$ of $\mathscr{D}$ having measure ε, where ε is small, and then Schwarz' inequality shows that $f^Y(z,s_n) \to 0$ uniformly in n as z tends to the cusp. Thus, $f^Y(z,s_n)$ is bounded on $\mathscr{D}$ for all n. (For a graphic indication of these arguments, see Fig. 6. The circle in Fig. 6 is $\{z' | d(z, z') = C\}$. The integral of $f^Y(z, s_n)$ on the interior of the circle reduces to the integral over the shaded area.) The assertion about strong convergence is now obvious. *q.e.d.*

An automorphic function which is an eigenfunction of the Laplacian D will be called a *cusp form* if it is square integrable on the fundamental domain and if the constant term of the Fourier expansion is 0 for every cusp. A cusp form is an automorphic function of the type described in Theorem 3.2.1, and is bounded on the upper half plane.

THEOREM 4.1.2. Let $f(z, s)$, $c_i(s)$ and $c_i'(s)$ be as in (4.1.1) and (4.1.2), and let $\{s_n\}$ be a convergent sequence of complex numbers such that $f(z, s_n)$ is defined for every n, and such that there are at most a finite number of n with $s_n(s_n - 1) = \lambda$ for any complex number λ. Assume that the norm $||f^Y(z, s_n)||$ of the compact part of f is bounded by a single constant for all n, that $\lim c_i(s_n) = \lim c_i'(s_n) = 0$ for all i, and that $f(z, s_n)$ has a pointwise limit $f(z)$. Then, $f(z) = 0$.

Proof. Since $f(z) = \lim f^Y(z, s_n)$, and since this convergence is strong by Theorem 4.1.1, $f(z)$ is a cusp form. On the other hand, almost all $f^Y(z,s_n)$ are orthogonal to a given cusp form because of Theorem 2.3.1. Therefore, $f(z)$ is orthogonal to all cusp forms, while $f(z)$ itself is a cusp form. This means $f(z) = 0$. *q.e.d.*

THEOREM 4.1.3. Let $f(z, s)$ and $c_i(s)$, $c_i'(s)$ be as in (4.1.1) and (4.1.2), and let $\{s_n\}$ be a convergent sequence of complex numbers as in Theorem 4.1.2. Furthermore, suppose that $c_i(s_n)$ and $c_i'(s_n)$ are bounded for all n and i. Then, the norm of the compact part $f^Y(z, s_n)$ is bounded by one constant for all n.

Proof. If $||f^Y(z, s_n)|| = a_n$ is not bounded, then Theorem 4.1.1 implies that there exists a subsequence $\{s_{n'}\}$ of $\{s_n\}$ such that $a_{n'} \to \infty$ and such that $\frac{1}{a_{n'}} f^Y(z, s_{n'})$ converges strongly to the compact part $f^Y(z)$ of some automorphic function $f(z)$. But, it follows from Theorem 4.1.2 that $f(z)$ is 0, while $||f^Y(z)|| = 1$. This is a contradiction. *q.e.d.*

THEOREM 4.1.4. Let $f(z, s)$, $c_i(s)$ and $c_i'(s)$ be as in (4.1.1) and (4.1.2), and let $\{s_n\}$ be a sequence of complex numbers as in Theorem 4.1.2 converging to $s_0 \in \boldsymbol{C}$. Furthermore, suppose that $c_i(s_n)$ and $c_i'(s_n)$ are bounded for all n and i. Then, there exists a subsequence of $\{f(z, s_n)\}$ converging pointwise and strongly to an automorphic function $f(z)$ with $Df = \lambda_0 f$, $(\lambda_0 = s_0(s_0 - 1))$.

Proof. By Theorem 4.1.3, $||f^Y(z, s_n)||$ is bounded. So, the assertion of the theorem follows from Theorem 4.1.1. *q.e.d.*

THEOREM 4.1.5. Let $f(z, s)$, $c_i(s)$ and $c_i'(s)$ be as in (4.1.1.) and (4.1.2), and let $\{s_n\}$ be a converging sequence of complex numbers such that $f(z, s_n)$ is defined for every n. Assume furthermore the norm of the compact part $f^Y(z, s_n)$ is bounded by a single constant for all n. Then, $c_i(s_n)$ and $c_i'(s_n)$ are bounded for all n and i.

Proof. Assume the contrary. Choosing, if necessary, a subsequence, we may assume that $|c_i(s_n)| \to \infty$ or $|c_i'(s_n)| \to \infty$, (as $n \to \infty$), for some i. It follows, however, from Theorem 4.1.1 that there exists a subsequence of $\{f(z, s_n)\}$ converging to an automorphic function. This is impossible. *q.e.d.*

§4.2. Continuation to Re $s = \frac{1}{2}$, $s \neq \frac{1}{2}$

In this and in the next article, we apply the results of §4.1 to the case in which $f(z, s)$ is specialized to an Eisenstein series $E_i(z, s)$ for a cusp κ_i belonging to a complete set $\kappa_1, \kappa_2, \dots, \kappa_h$ of inequivalent cusps of a discontinuous group Γ of finite type. If we are studying only one Eisenstein series $E_1(z, s)$, then it will be denoted by $E(z, s)$.

THEOREM 4.2.1. Let Γ be a discontinuous group of finite type; then every entry $\varphi_{ij}(s)$ of the constant term matrix $\Phi(s)$ is bounded as $s \to s_0 = \frac{1}{2} + \sqrt{-1}t_0$, $(t_0 \neq 0)$.

Proof. By Theorem 4.1.5, it is enough to show that the norm of the compact part of an Eisenstein series is bounded as $s \to s_0$, and we may reduce the proof to showing the boundedness of $||E^Y(z, s)||$ as $s \to s_0$, for one Eisenstein series $E(z, s) = E_1(z, s)$ for κ_1, where $\kappa_1, \kappa_2, \dots, \kappa_h$ form a complete set of inequivalent cusps of Γ.

If $||E^Y(z, s)||$ is not bounded as $s \to s_0$, then it follows from Theorem 4.1.1 that there exists a sequence $\{s_n\}$ of complex numbers tending to s_0 such that

$$g^Y(z) = \lim \frac{1}{a_n} E^Y(z, s_n), \qquad (a_n = ||E^Y(z, s_n)||),$$

is the compact part of an automorphic function $g(z)$. The constant term of the Fourier expansion of $g(z)$ at κ_i is of the form $c_i' y^{1-s_0}$, and Theorem 4.1.2 shows that at least one c_i' is not 0. It follows now from Theorem 2.3.1 that

$$\left(g^Y(z), \frac{1}{a_n} E^Y(z,s_n)\right) = \sum_{i=1}^{h} \frac{1}{a_n} \left(\frac{-c_i' \varphi_i(\bar{s}_n)}{s_0 + \bar{s}_n - 1} Y^{-s_0 - \bar{s}_n + 1} + \frac{-c_i' \delta_{1i}}{s_0 - \bar{s}_n} Y^{-s_0 + \bar{s}_n} \right).$$

The left hand side should be finite as $n \to \infty$, but on the right hand side we have $\varphi_i(s_n)/a_n \to c_i'$, $s_0 + s_n - 1 \to 0$ and $s_0 - \bar{s}_n \to 2\sqrt{-1}t_0 \neq 0$. So the right hand side becomes infinitely large. This is a contradiction. *q.e.d.*

THEOREM 4.2.2. Let Γ be a discontinuous group of finite type, and let $\{s_n\}$ be a sequence of complex numbers with Re $s_n > 1/2$ such that $\lim s_n = 1/2 + \sqrt{-1}t_0$, $(t_0 \neq 0)$, and such that the sequence $\Phi(s_n)$ of matrices converges. Then, $\lim \Phi(s_n)$ is a unitary matrix.

Proof. Denote by $\kappa_1, \kappa_2, \ldots, \kappa_h$ a complete set of inequivalent cusps of Γ, and let $E_i(z, s)$ be the Eisenstein series for κ_i. Then, by Theorem 4.1.3, the norm of the compact part $E_i^Y(z, s)$ is bounded for all i and n. On the other hand, Theorem 2.3.2 shows

$$(E_i^Y(z, s_n), E_j^Y(z, \bar{s}_n)) = \frac{\delta_{ij} Y^{s_n + \bar{s}_n - 1} - \sum_{l=1}^{h} \varphi_{il}(s_n) \varphi_{jl}(\bar{s}_n) Y^{-s_n - \bar{s}_n + 1}}{s_n + \bar{s}_n - 1}$$

$$+ \varphi_{ji}(\bar{s}_n) \frac{Y^{s_n - \bar{s}_n}}{s_n - \bar{s}_n} - \varphi_{ij}(s_n) \frac{Y^{-s_n + \bar{s}_n}}{s_n - \bar{s}_n}, \qquad ((\varphi_{ij}(s)) = \Phi(s)).$$

The left hand side is bounded. Therefore, we have

$$\lim \sum_{l=1}^{h} \varphi_{il}(s_n) \varphi_{jl}(\bar{s}_n) = \delta_{ij}$$

on the right land side. Since the constant term matrix is real for real values of s, we have $\varphi_{ij}(\bar{s}) = \overline{\varphi_{ij}(s)}$. *q.e.d.*

THEOREM 4.2.3. Let Γ be a discontinuous group of finite type, and let $s_0 = 1/2 + \sqrt{-1}t_0$ be a complex number with Im $s_0 = t_0 \neq 0$. Then, there exists a unique unitary matrix $\Phi(s_0)$ such that $\lim \Phi(s_n) = \Phi(s_0)$ for any sequence $\{s_n\}$ with $\mathrm{Re}\, s_n > 1/2$, $\mathrm{Im}\, s_n \neq 0$, tending to s_0.

Proof. Assume that there exist two sequences $\{s_n\}, \{s_n'\}$ giving different limits for Φ. Put $\lim \Phi(s_n') = (c_{ij}')$. By Theorem 4.1.4 there exists an automorphic function $g(z)$ such that $\lim E^Y(z, s_n') = g^Y(z)$ holds in the sense of pointwise convergence and in the sense of strong convergence. Here, $E(z,s)$ means $E_1(z, s)$ for an element κ_1 of a complete set $\kappa_1, \kappa_2, \dots, \kappa_h$ of inequivalent cusps; however, κ_1 is not necessarily ∞. By Theorem 2.3.1, we have now

$$(g^Y(z), E^Y(z, s_n)) = \frac{Y^{s_0+\bar{s}_n-1} - \sum_{i=1}^{h} c_i' \varphi_i(\bar{s}_n) Y^{-s_0-\bar{s}_n+1}}{s_0 + \bar{s}_n - 1} + \frac{\varphi_1(\bar{s}_n) Y^{s_0-\bar{s}_n} - c_1' Y^{-s_0+\bar{s}_n}}{s_0 - \bar{s}_n}$$

with $c_{1i} = c_i$. The left hand side is bounded by Theorem 4.1.3, but the right hand side is not bounded by Theorem 4.2.2 under the given assumptions. This is a contradiction. *q.e.d.*

§4.3. Continuation to $[\frac{1}{2}, 1]$

THEOREM 4.3.1. Let Γ be a discontinuous group of finite type, and let $S_0 > 1/2$ be a real number. Then, there exists a unique matrix $\Phi(S_0)$ such that $\lim \Phi(s_n) = \Phi(S_0)$ for any sequence $\{s_n\}$ with $\mathrm{Re}\, s_n > 1/2$, $\mathrm{Im}\, s_n \neq 0$, tending to S_0. Here, ∞ is allowed as an entry of $\Phi(S_0)$.

Proof. To prove this theorem, we assume first that Γ is reduced at ∞, and show that $\varphi(s)$, defined as in (2.2.7), is meromorphic on the interval $(1/2, 1]$.

Let $\kappa_1 = \infty, \kappa_2, \dots, \kappa_h$ be a complete set of inequivalent cusps of Γ, and denote, in general, by $f(z, s)$, $g(z, s')$ automorphic functions chosen as in Theorem 2.3.1 such that $c_2 = \dots = c_h = 0$, $d_2 = \dots = d_h = 0$. Furthermore, denote by $\tilde{f}^Y(z, s)$, $\tilde{g}^Y(z, s')$ functions defined in the same way as $\tilde{E}^Y(z, s)$ was defined in Theorem 2.3.3. Then, by substituting in the formula of Theorem 2.3.3, we have the inner product formula

$$(4.3.1) \qquad (\tilde{f}^Y(z, s), \overline{\tilde{g}^Y(z, s')}) = \frac{c_1 d_1 Y^{s+s'-1}}{s + s' - 1} + \frac{c_1 d_1' Y^{s-s'} - c_1' d_1 Y^{-s+s'}}{s - s'},$$

provided that $\mathrm{Re}\, s > 1/2$, $\mathrm{Re}\, s' > 1/2$.

Using this formula, we prove here that $||E^Y(z, s_n)|| \to \infty$ if and only if $|\varphi(s_n)| \to \infty$ for any sequence $s_n \to S_0$. It is clear by Theorem 4.1.5 that

$||E^Y(z, s_n)|| \to \infty$ if $|\varphi(s_n)| \to \infty$. Suppose therefore that there exists a sequence $\{s_n\}$ for which we have $a_n = ||E^Y(z, s_n)|| \to \infty$ while $|\varphi(s_n)|$ is bounded. Then, by Theorem 4.1.1, we can find a subsequence of $\frac{1}{a_n} E(z, s_n)$ having a limit $g(z)$ which is not zero. However, the constant term of the Fourier expansion of $g(z)$ is 0 by assumption. So, (4.3.1) yields

$$\left(\tilde{g}^Y(z), \frac{1}{a_n} \tilde{E}^Y(z, s)\right) = 0,$$

which is a contradiction. This shows at the same time that $\lim\varphi(s_n)/a_n \neq 0$, whenever $a_n \to \infty$.

Assume now there exist two sequences $\{s_n\}$, $\{s_n'\}$ converging to S_0 such that $\lim\varphi(s_n)$ and $\lim \varphi(s_n')$ are finite but different; put $\lim\varphi(s_n') = c'$ and find a subsequence of $E(z, s_n')$ having a limit $g(z)$ by means of Theorem 4.1.4. Then, it follows from (4.3.1) that

$$(\tilde{g}^Y(z), \tilde{E}^Y(z, s_n)) = \frac{Y^{S_0+\bar{s}_n-1}}{S_0 + \bar{s}_n - 1} + \frac{\varphi(\bar{s}_n)Y^{S_0-\bar{s}_n} - c'Y^{-S_0+\bar{s}_n}}{S_0 - \bar{s}_n}$$

This leads to a contradiction when we let $n \to \infty$.

Assume next that $\lim\varphi(s_n)$ is finite and that $\lim|\varphi(s_n')| = \infty$. Then, $||\tilde{E}^Y(z, s_n')|| = a_n' \to \infty$ as $n \to \infty$ by Theorem 4.1.5, and, by Theorem 4.1.4, we can find a subsequence of $\frac{1}{a_n'}E(z, s_n')$ having a limit $g(z)$ which is different from 0. But, (4.3.1) shows in this case that

$$(\tilde{g}^Y(z), \tilde{E}^Y(z, s_n)) = \frac{-c'Y^{-S_0+\bar{s}_n}}{S_0 - \bar{s}_n},$$

where c' is a constant such that $c'Y^{1-S_0}$ is the constant term of the Fourier expansion of $g(z)$ at ∞. Since $c' \neq 0$, as shown above, we have a contradiction. This proves our assertion on $\varphi(s)$.

In the general case, consider for each cusp κ_i a function $\tilde{E}_i^Y$ which is, similarly to $\tilde{E}^Y$ above, obtained from the Eisenstein series $E_i(z, s)$ for κ_i by cutting off the first term in the constant term (2.2.9) of the Fourier expansion of E_i at κ_i, and investigate the inner product $(\tilde{E}_i^Y(z, s), \tilde{E}_j^Y(z, \overline{s'}))$ which is equal to

$$\frac{\delta_{ij}Y^{s+s'-1}}{s + s' - 1} + \varphi_{ji}(s')\frac{Y^{s-s'}}{s - s'} - \varphi_{ij}(s)\frac{Y^{-s+s'}}{s - s'}$$

by Theorem 2.3.2. Then, the same type of arguments as developed above will lead to the conclusion that $\varphi_{ii}(s)$ as well as $\varphi_{ij}(s)$ is meromorphic on the interval (½, 1]. *q.e.d.*

This theorem assures that the function φ_{ij} is meromorphic even on the interval $(\frac{1}{2}, 1]$. Therefore, $\varphi_{ij}(s)$ is holomorphic in $\operatorname{Re} s > \frac{1}{2}$ except for poles on $(\frac{1}{2}, 1]$ which may accumulate, if they do, only at $\frac{1}{2}$. The following theorems show, however, that such poles are all simple, and finite in number.

THEOREM 4.3.2. Let $\Phi(s) = (\varphi_{ij}(s))$ be the constant term matrix of Eisenstein series for a discontinuous group of finite type. Then, every pole of $\varphi_{ij}(s)$ at a point on $(\frac{1}{2}, 1]$ is simple.

Proof. It is enough to show that $\varphi_{ij}(S_0 + \sqrt{-1}t) = O(t^{-1})$ as $t \to 0$ for $\frac{1}{2} < S_0$. This follows from the integral representation (3.1.6.) because

$$||R_\lambda|| \leqq [\operatorname{Im} \zeta(\zeta - 1)]^{-1}$$

by (3.1.5), and $[\operatorname{Im}\zeta(\zeta - 1)]^{-1} = O(t^{-1})$ for $\zeta = S_0 + \sqrt{-1}t$. *q.e.d.*

THEOREM 4.3.3. Let $\Phi(s) = (\varphi_{ij}(s))$ be the constant term matrix of Eisenstein series for a discontinuous group Γ of finite type. Then, $\Phi(s_n)$ has a unique limit for any sequence $\{s_n\}$ of complex numbers such that $\operatorname{Re} s_n > \frac{1}{2}$, $\lim s_n = \frac{1}{2}$.

Proof. We write $E(z, s)$ for $E_1(z, s)$ and $\varphi_l(s)$ for $\varphi_{1l}(s)$; it is enough to prove the theorem for φ_l.

For two sequences $\{s_n\}$, $\{s_n'\}$ converging to $\frac{1}{2}$, put

$$\zeta = \zeta_{n,n'} = \left(s_n - \frac{1}{2}\right) + \left(s'_{n'} - \frac{1}{2}\right),$$

$$\zeta' = \zeta'_{n,n'} = \left(s_n - \frac{1}{2}\right) - \left(s'_{n'} - \frac{1}{2}\right).$$

Then, it follows from Theorem 2.3.2 that

(4.3.2)

$$(E^Y(z, s_n), \overline{E^Y(z, s'_{n'})}) = \frac{1}{\zeta}\left(Y^\zeta - \sum_{l=1}^{h} \varphi_l\left(\frac{\zeta+\zeta'}{2} + \frac{1}{2}\right)\varphi_l\left(\frac{\zeta-\zeta'}{2} + \frac{1}{2}\right)Y^{-\zeta}\right)$$
$$+ \frac{1}{\zeta'}\left(\varphi_1\left(\frac{\zeta-\zeta'}{2} + \frac{1}{2}\right)Y^{\zeta'} - \varphi_1\left(\frac{\zeta+\zeta'}{2} + \frac{1}{2}\right)Y^{-\zeta'}\right).$$

Now, in general, if $f(\zeta, \zeta')$, $f'(\zeta, \zeta')$ are two functions of ζ, ζ' which are continuous at $(\zeta, \zeta') = (0, 0)$, and if

$$\frac{1}{\zeta} f(\zeta, \zeta') + \frac{1}{\zeta'} f'(\zeta, \zeta')$$

remains finite as ζ and ζ' tend to 0 independently, then we have $f(0, 0) = f'(0, 0) = 0$. Even if ζ, ζ' are restricted by the condition that (ζ, ζ')

should belong to a certain subset M of $C \times C$, we still obtain the same conclusion, whenever there exist in M two sequences $\{\eta_n, \eta_n'\}$ and $\{\zeta_n, \zeta_n'\}$ tending to (0, 0) such that the sequences $\{\eta_n/\eta_n'\}$ and $\{\zeta_n/\zeta_n'\}$ of the quotients of the coordinates of the points in the two sequences have different limits. Therefore, if $\lim \varphi_l(s_n)$ is finite for all l, then Theorem 4.1.4 and the above inner product formula (4.3.2) with $\{s_n'\} = \{s_n\}$ show that $\lim \sum |\varphi_l(s_n)|^2 = 1$. So, if there exist two sequences $\{s_n\}$, $\{s_n'\}$ such that $\lim \varphi_l(s_n)$ and $\lim \varphi_l(s_n')$ are finite but different for some l, then (4.3.2) will lead to a contradiction.

Assume next that there exists a sequence $\{s_n\}$ such that $\lim|\varphi_l(s_n)| = \infty$ for some l. Then, since $||E^Y(z, s_n)|| = a_n \to \infty$ by Theorem 4.1.5, the formula for the inner product $(a_n^{-1}E^Y(z, s_n), \overline{a_n^{-1}E^Y(z, s_n)})$ derived from (4.3.2) will lead to a contradiction. *q.e.d.*

Theorem 3.1.4, Theorem 4.2.2, Theorem 4.3.1, Theorem 4.3.2, and Theorem 4.3.3 imply now

THEOREM 4.3.4. Let $\Phi(s) = (\varphi_{ij}(s))$ be the constant term matrix of the Eisenstein series for a discontinuous group of finite type. Then, every $\varphi_{ij}(s)$ is a meromorphic function in the domain $\text{Re } s > \frac{1}{2}$, has at most a finite number of simple poles on the interval $(\frac{1}{2}, 1]$, and $\Phi(s)$ converges to a unique unitary matrix $\Phi(s_0)$ when s tends to a point s_0 on the line $\text{Re } s = \frac{1}{2}$.

This theorem, together with Theorem 3.2.2, Theorem 4.1.2, and Theorem 4.1.4, yields the following important theorem on the analytic continuation of Eisenstein series.

THEOREM 4.3.5. Let Γ be a discontinuous group of finite type, and let $E_i(z, s)$ be the Eisenstein series for a cusp of Γ. Denote by $\Phi(s) = (\varphi_{ij}(s))$ the constant term matrix. Then, $E_i(z, s)$ is holomorphic in the domain $\text{Re } s > \frac{1}{2}$ except at a finite number of poles of $\varphi_{ii}(s)$ on $(\frac{1}{2}, 1]$. Moreover, at a pole of φ_{ii}, the function $\varphi_{ii}(s)^{-1}E_i(z, s)$ is holomorphic. Finally, $E_i(z, s)$ has a unique limit as s tends to a point s_0 on the line $\text{Re } s = \frac{1}{2}$.

§4.4. Continuation to the whole plane and the functional equation

What we have proved in §4.3 yields now at once complete results on the analytic continuation of Eisenstein series.

THEOREM 4.4.1. Let $\Phi(s)$ be the constant term matrix of an Eisenstein series for a discontinuous group of finite type. Then, $\Phi(s)$ is meromorphic on the whole s-plane, and satisfies the functional equation $\Phi(s)\Phi(1-s) = I$, where I is the identity matrix.

Proof. Let $s_0 = \frac{1}{2} + \sqrt{-1}t_0$ be a point on the line $\operatorname{Re} s = \frac{1}{2}$, and denote by $\Phi(s_0)$ the unique limit given by Theorem 4.3.4 as $s \to s_0$, ($\operatorname{Re} s > \frac{1}{2}$). Then, $\Phi(s_0)$ is unitary, and it is also symmetric by Theorem 2.2.1. So, we have

$$\Phi(s_0)\Phi(1-s_0) = \Phi(s_0)\Phi(\bar{s}_0) = \Phi(s_0)\overline{{}^t\Phi(s_0)} = I,$$

where t on the left shoulder of a matrix denotes transposition. This proves the functional equation on the line $\operatorname{Re} s = \frac{1}{2}$. Therefore, the theorem follows from the reflection principle in function theory. *q.e.d.*

THEOREM 4.4.2. Let Γ be a discontinuous group of finite type, let $E_i(z,s)$, ($i = 1, 2, \ldots, h$), be the Eisenstein series for a complete set $\kappa_1, \kappa_2, \ldots, \kappa_h$ of inequivalent cusps of Γ, and let $\Phi(s)$ be the constant term matrix of $E_i(z, s)$. Then, each $E_i(z, s)$ has a meromorphic continuation to the whole s-plane, and the column vector $\mathscr{E}(z, s) = {}^t(E_1, \ldots, E_h)$ satisfies the functional equation

$$\mathscr{E}(z, s) = \Phi(s)\mathscr{E}(z, 1-s).$$

Proof. Let s_0 be on the line $\operatorname{Re} s = \frac{1}{2}$, and denote by $\mathscr{E}(z,s_0)$ the unique limit given by Theorem 4.3.5 as $s \to s_0$, ($\operatorname{Re} s > \frac{1}{2}$). Furthermore, denote by $A = (a_{ij})$ the matrix such that a_{ij} is equal to the constant term in the Fourier expansion at κ_j of the i-th component of the vector $\mathscr{E}(z, s_0) - \Phi(s_0)\mathscr{E}(z, 1-s_0)$; $\Phi(s_0)$ is certainly well-defined by Theorem 4.4.1. Then, it follows from Theorem 4.4.1 that

$$A = Iy^{s_0} + \Phi(s_0)y^{1-s_0} - \Phi(s_0)(Iy^{1-s_0} + \Phi(1-s_0)y^{s_0}) = 0.$$

Therefore, Theorem 4.1.2 shows $\mathscr{E}(z, s_0) - \Phi(s_0)\mathscr{E}(z, 1-s_0) = 0$. This is the functional equation on the line $\operatorname{Re} s = \frac{1}{2}$. So, using the reflection principle, we get the desired conclusion. *q.e.d.*

The rest of this § is devoted to miscellaneous remarks on Eisenstein series. First we propose to explain a certain geometric meaning of the Dirichlet series $\varphi_{ij}(s)$ in (2.2.10) appearing in the constant term of the Fourier expansion of an Eisenstein series, and at the same time to verify the divergence of the Eisenstein series for $\operatorname{Re} s \leqq 1$.† For the sake of simplicity, we consider a discontinuous group Γ of finite type reduced at ∞, and the Eisenstein series $E(z, s) = E_1(z, s)$ for the cusp $x_1 = \infty$. Denote by $\mathscr{D}_Y{}'$

† Dr. A. F. Beardon has communicated simple proofs for the corresponding divergence problem for a discontinuous group acting on the unit circle.

the domain in H determined by $\operatorname{Im} z > Y$, $0 < \operatorname{Re} z < 1$, and let $\mathscr{D}$ be a connected fundamental domain containing $\mathscr{D}_Y'$. Furthermore, let κ be a cusp on $\boldsymbol{R}$ equivalent to ∞, and let $\Gamma_{\infty,\kappa}$ be the set of all $\sigma \in \Gamma$ with $\sigma\infty = \kappa$. Then, for every $\sigma = \begin{pmatrix} a & b \\ c & d \end{pmatrix} \in \Gamma_{\infty,\kappa}$, the absolute value $|c|$ is the same, and $\bigcup \sigma \mathscr{D}_Y', (\sigma \in \Gamma_{\infty,\kappa})$, becomes a euclidean circle with radius $1/2c^2 Y$ tangential to the real axis at κ. As one sees by an elementary calculation, the integral I_κ of $y(z)dz$ over the interior of the circle is equal to $\pi/c^2 Y$. Therefore, denoting $\varphi_{11,0}$ (resp. by $\varphi_{1j,0}$), as defined in (2.2.10), simply by φ_0 (resp. by $\varphi_{j,0}$), the series $\varphi_0(1)$ is formally equal to the sum of all I_κ with $0 \leqq \kappa < 1$ multiplied by the constant π/Y. Let κ_j be a cusp on the boundary of $\mathscr{D}$ different from ∞. Then, for any cusp κ equivalent to κ_j, we may consider similarly a circle which is the union of all $\sigma\sigma_j \mathscr{D}_Y'$, where σ_j is as in the definition (2.1.1) of Eisenstein series, and σ runs over the set of all $\sigma \in \Gamma$ such that $\sigma\kappa_j = \kappa$. The integral $I_{j,\kappa}$ of $y(z)dz$ over the interior of the circle produces in this case a term in the formal series $\varphi_{j,0}(1)$ multiplied by π/Y. Assume now $\varphi_0(1)$ to be finite; then, Theorem 4.3.5 implies that all other $\varphi_{j,0}(1)$ are finite. Therefore, the integral of $y(z)dz$ over the union of the domains $\sigma(\mathscr{D} - \bigcup_j \sigma_j \mathscr{D}_Y')$, $(\sigma \in \Gamma_0 \backslash \Gamma)$, is infinite. Using this fact, however, one can argue exactly as in the proof of Theorem 2.1.1 taking a point $z \in \mathscr{D} - \bigcup_j \sigma_j \mathscr{D}_Y'$ and drawing a sufficiently large non-euclidean circle with center z, and can conclude that the sum of all $y(\sigma z)$, $(\sigma \in \Gamma_0 \backslash \Gamma)$, is infinite. If $\sum y(\sigma z)$ is infinite, then $\varphi_0(1)$ is also infinite by Theorem 4.3.5, so that we have a contradiction. Thus, $s = 1$ is always a pole of $E(z, s)$, and consequently the series $E(z, s)$ does not converge when $\operatorname{Re} s \leqq 1$.

Next some comments will be added concerning the special case where the discontinuous group Γ is equal to the elliptic modular group $SL(2,\boldsymbol{Z})$. We can take as a fundamental domain in this case the domain $\mathscr{D} = \{z \in H \mid |z| > 1, -\frac{1}{2} < \operatorname{Re} z < \frac{1}{2}\}$; there is only one cusp ∞ of $\mathscr{D}$, so there is only one Eisenstein series $E(z, s)$ for Γ given by

$$E(z, s) = \frac{1}{2}\sum \frac{y^s}{|cz + d|^{2s}}, \qquad (c, d \in \boldsymbol{Z},\ (c, d) = 1),$$

where (c, d) means the greatest common divisor of c and d. Let us now try to find some explicit formulas for the coefficients $a_m(y, s)$ in the Fourier expansion

$$E(z, s) = \sum_{m=-\infty}^{\infty} a_m(y, s) e(mx).$$

By (2.2.4), (2.2.6), and (2.2.7), it is easy to see that

$$a_0(y, s) = y^s + \varphi(s)y^{1-s}$$

with

$$\varphi(s) = \pi^{\frac{1}{2}} \frac{\Gamma(s - \frac{1}{2})}{\Gamma(s)} \varphi_0(s)$$

where

$$\varphi_0(s) = \zeta(2s - 1) / \zeta(2s), \tag{4.4.1}$$

and $\zeta(s)$ is the Riemann zeta function. By (2.2.5), we have furthermore

$$a_m(y, s) = 2\pi^s |m|^{s-\frac{1}{2}} \Gamma(s)^{-1} y^{\frac{1}{2}} K_{s-\frac{1}{2}}(2\pi|m|y)\varphi_m(s) \tag{4.4.2}$$

for $m \neq 0$, and it follows from (2.2.4) and from the property

$$\mu(c) = \sum e\left(\frac{d}{c}\right), \qquad (d \bmod c, \quad (d, c) = 1),$$

of the Möbius function $\mu(c)$ that

$$\varphi_m(s) = \zeta(2s)^{-1} \cdot \prod_{p \mid m} \frac{1 - p^{-(e+1)(2s-1)}}{1 - p^{-2s+1}} = \zeta(2s)^{-1} \sum_{d \mid m} d^{-2s+1}, \tag{4.4.3}$$

where e is the p-exponent of m.

A remarkable consequence of (4.4.1) is that $\varphi(\frac{1}{2}) = -1$. This means that the constant term matrix $\Phi(s)$ of the Eisenstein series is not necessarily the identity matrix even at $s = \frac{1}{2}$, while $\Phi(\frac{1}{2})$ is a real, symmetric, and unitary matrix.

Moreover, it should be noted that the theory of analytic continuation of the Eisenstein series furnishes an analytic continuation of the Riemann zeta function in virtue of (4.4.1). It is also possible to deduce the functional equation of the Riemann zeta function by means of the functional equation of the Eisenstein series. To see this, consider the function $a_1(y, s)$ appearing in (4.4.2); then Theorem 4.4.2 yields

$$a_1(y, s) = \pi^{\frac{1}{2}} \frac{\Gamma(s - \frac{1}{2})}{\Gamma(s)} \frac{\zeta(2s - 1)}{\zeta(2s)} a_1(y, 1 - s).$$

Then, using (4.4.3) as well as the relation $K_s = K_{-s}$, and putting $s = \frac{1}{2}(1 + s')$, we have the functional equation

$$\pi^{-\frac{s'}{2}} \Gamma\left(\frac{s'}{2}\right) \zeta(s') = \pi^{-\frac{1-s'}{2}} \Gamma\left(\frac{1 - s'}{2}\right) \zeta(1 - s')$$

for $\zeta(s)$. These results on the Riemann zeta function are usually based upon the Poisson summation formula for the Fourier transformation, but the theory of Eisenstein series affords an entirely different proof of them.

CHAPTER V

Spectral theory of $L^2(\Gamma\backslash H)$

§5.1. Cusp forms and incomplete theta series

The Hilbert space $\mathfrak{H} = L^2(\Gamma\backslash H)$ of square integrable automorphic functions with respect to a discontinuous group Γ of finite type and with respect to the inner product (2.3.1) is a countably infinite dimensional linear space over $\boldsymbol{C}$. In spite of its simple abstract structure, $\mathfrak{H}$ is a very important example of a space in which the Laplacian $D = y^2\left(\frac{\partial^2}{\partial x^2} + \frac{\partial^2}{\partial y^2}\right)$ of the upper half plane H has a number of spectra, both discrete and continuous. Moreover, the continuous spectrum can be expressed explicitly by using the values of Eisenstein series on the line $s = \frac{1}{2} + \sqrt{-1}t$, $(t \in \boldsymbol{R})$.

As a matter of fact, a wholly analogous situation exists for the Hilbert space $L^2(\Gamma\backslash G)$, with $G = SL(2, \boldsymbol{R})$, and the spectral decomposition of $L^2(\Gamma\backslash G)$ will lead to the decomposition of a certain natural unitary representation into its irreducible components. Since the theory for $L^2(\Gamma\backslash G)$ requires still more preparation, however, it must wait until Chap. VII. In this chapter, we undertake the study of the simpler space $L^2(\Gamma\backslash H)$ as a step preliminary to the consideration of $L^2(\Gamma\backslash G)$. In fact, $\Gamma\backslash H = \Gamma\backslash G/K$ with $K = SO(2)$.

Denote now by $\kappa_1, \kappa_2, \dots, \kappa_h$ a complete set of inequivalent cusps, and by σ_i an element in $SL(2, \boldsymbol{R})$ such that $\sigma_i \infty = \kappa_i$, and such that $\sigma_i^{-1}\Gamma_i\sigma_i = \Gamma_0 = \left\{\begin{pmatrix} 1 & b \\ & 1 \end{pmatrix} \middle| b \in \boldsymbol{Z}\right\}$, where $\Gamma_i = \{\sigma \in \Gamma | \sigma\, \kappa_i = \kappa_i\}$ is the stabilizer of κ_i in Γ. Let N be the group of all $\begin{pmatrix} 1 & x \\ & 1 \end{pmatrix}$, $(x \in \boldsymbol{R})$. Then, by (3.1.3), a function ψ on $N\backslash H$ determines an automorphic function $\theta_{i,\psi}(z)$. From now on this function will be called an *incomplete theta series*. The function ψ may, of course, be regarded as a function of $y > 0$, and the function ψ in an incomplete theta series is always assumed to decrease sufficiently rapidly both as $y \to 0$ and as $y \to \infty$ in order that $\theta_{i,\psi}$ belong to $\mathfrak{H} = L^2$

$(\Gamma\backslash H)$. The smallest closed subspace of $\mathfrak{H}$ containing the span of all the $\theta_{i,\psi}$ will be denoted by Θ. If one considers the closed subspace of $\mathfrak{H}$ spanned by all the $\theta_{i,\psi}(z)$ with certain restrictions on ψ, e.g. ψ differentiable or ψ of compact support, the resulting space Θ will be the same.

The smallest closed subspace of $\mathfrak{H}$ containing the cusp forms with respect to Γ will be denoted by $\mathfrak{H}_0$. The space of automorphic functions $f \in \mathfrak{H}$ such that the constant term in the Fourier expansion of f at κ_i is 0 for all i will be denoted by $\mathfrak{H}_0'$. It is clear that $\mathfrak{H}_0' \supset \mathfrak{H}_0$, but it is shown in the next § that actually $\mathfrak{H}_0' = \mathfrak{H}_0$ holds.

THEOREM 5.1.1. Let Γ be a discontinuous group of finite type. Then, the orthogonal complement in $\mathfrak{H} = L^2(\Gamma\backslash H)$ of the space Θ of incomplete theta series is the space $\mathfrak{H}_0'$.

Proof. Let the definitions of κ_i, Γ_i, and σ_i be as explained above, and let $\theta_{i,\psi} \in \Theta, f \in \mathfrak{H}$. Then,

$$(\theta_{i,\psi}, f) = \int_{\Gamma\backslash H} \theta_{i,\psi}(z)\overline{f(z)}dz = \int_{\Gamma_i\backslash H} \psi(\sigma_i^{-1}z)\overline{f(z)}dz$$

$$= \int_{\Gamma_0\backslash H} \psi(z)\overline{f(\sigma_i z)}dz = \int_0^\infty \psi(y) \int_0^1 f(\sigma_i z)dx \frac{dy}{y^2}.$$

Therefore, in order that $(\theta_{i,\psi}, f) = 0$ for all $\theta_{i,\psi} \in \Theta$, it is necessary and sufficient that

$$\int_0^1 f(\sigma_i z)dx = 0$$

for all i. *q.e.d.*

It is easy to see that both Θ and $\mathfrak{H}_0$ are infinite-dimensional. We shall show in the next § that the Laplacian $D = y^2\left(\frac{\partial^2}{\partial x^2} + \frac{\partial^2}{\partial y^2}\right)$ has only a discrete spectrum in $\mathfrak{H}_0'$ $(= \mathfrak{H}_0)$.

Denote now by $F(z)$ a function on the upper half plane H such that F may be regarded as a function on $K\backslash G/K$. As a function of z, $F(z)$ is invariant under all the transformations $z \to \sigma z$ with $\sigma \in K = SO(2)$. Let $k(z, z')$ be the point pair invariant determined by F, i.e., if $z = g \cdot \sqrt{-1}$, $z' = g' \cdot \sqrt{-1}$, $(g, g' \in G)$, put $k(z, z') = F(g'^{-1}g)$. For the invariant integral operator determined by $k(z, z')$, we have then

$$\int_H k(z, z')f(z')dz' = \int_{\Gamma\backslash H} \sum_{\sigma\in\Gamma} k(z, \sigma z')f(z')dz',$$

whenever f is an automorphic function with respect to the discontinuous group Γ. In other words, if f is automorphic, the integral operator determined by $k(z, z')$ has the same effect on f as the integral operator

$$f \rightarrow \int_{\mathscr{D}} K(z, z')f(z')dz' \tag{5.1.1}$$

on the fundamental domain, where

$$K(z, z') = \sum_{\sigma \in \Gamma} k(z, \sigma z'). \tag{5.1.2}$$

By Theorem 1.3.2 and Theorem 1.3.3, the discrete spectrum of the Laplacian is completely determined if we know the spectrum of every integral operator of the form (5.1.1) whose kernel function K is defined by a function on $K\backslash G/K$ which is C^∞ and of compact support.

A basic technique for the study of the spectrum of such an integral operator is given by the following theorem which shows how to separate cusp forms from other automorphic functions.

THEOREM 5.1.2. Let Γ be a discontinuous group of finite type, let $\kappa_1, \kappa_2, \ldots, \kappa_h$ be a complete set of inequivalent cusps of Γ, and let Γ_i be the stabilizer in Γ of κ_i. Moreover, denote by Θ the space of incomplete theta series, and let $\mathfrak{H} = \Theta \oplus \mathfrak{H}_0'$ be the orthogonal decomposition of $\mathfrak{H} = L^2(\Gamma\backslash H)$. Furthermore, for a point pair invariant $k(z, z')$, put

$$H(z, z') = \sum_{i=1}^{h} H_i(z, z'),$$

$$H_i(z, z') = \sum_{\sigma} \int_{-\infty}^{\infty} k(z, \sigma_i n(x) \sigma_i^{-1} \cdot \sigma z')dx, \qquad (\sigma \in \Gamma_i\backslash\Gamma),$$

where σ_i is a linear transformation in G such that $\sigma_i^{-1}\Gamma_i\sigma_i = \Gamma_0 = \{n(b) | b \in \boldsymbol{Z}\}$, and $n(x) = \begin{pmatrix} 1 & x \\ & 1 \end{pmatrix}$. Then, we have

$$\int_{\mathscr{D}} H(z, z')f(z')dz' = 0, \qquad (\mathscr{D} = \Gamma\backslash H),$$

for any $f \in \mathfrak{H}_0'$.

Proof. Let N be the group of all $n(x)$, $x \in R$. Then,

$$\int_{\mathscr{D}} H_i(z, z')f(z')dz' = \int_{\mathscr{D}} \Big[\sum_{\sigma} \int_{-\infty}^{\infty} k(z, \sigma_i\, n(x)\sigma_i^{-1} \cdot \sigma z')dx \Big] f(z')dz'$$

$$= \int_{\Gamma_i\backslash H}\left[\int_{-\infty}^{\infty} k(z, \sigma_i\, n(x)\sigma_i^{-1}\, z')dx\right] f(z')dz'$$

$$= \int_{\Gamma_0\backslash H}\left[\int_{-\infty}^{\infty} k(z, \sigma_i\, n(x)z')dx\right] f(\sigma_i\, z')dz'$$

$$= \int_{N\backslash H}\int_{\Gamma_0\backslash N}\left[\int_{-\infty}^{\infty} k(z, \sigma_i\, n(x)n(x')\, a(y')\, \sqrt{-1})dx\right] \times f(\sigma_i\, n(x')\, a(y')\, \sqrt{-1})\, dx'\, \frac{dy'}{y'^2}$$

$$= \int_{N\backslash H}\left(\int_{-\infty}^{\infty} (k(z, \sigma_i\, n(x)\, a(y')\, \sqrt{-1})dx\right) \times \left(\int_{\Gamma_0\backslash N} f(\sigma_i\, n(x')\, a(y')\, \sqrt{-1})dx'\right)\frac{dy'}{y'^2} = 0$$

with $z' = x' + \sqrt{-1}y'$, $a(y') = \begin{pmatrix} \sqrt{y'} & \\ & \sqrt{y'^{-1}} \end{pmatrix}$. *q.e.d.*

This theorem shows that the spectrum of the integral operator (5.1.1) in $\mathfrak{H}_0'$ is exactly the same as the spectrum of $K(z, z') - H'(z, z')$.

THEOREM 5.1.3. Let $\mathfrak{H}_0'$ be as in Theorem 5.1.1. Then, the invariant operator defined by a point pair invariant $k(z, z')$ maps $\mathfrak{H}_0'$ into itself.

Proof. Using the same notation as in the preceding theorems of this §, we have $\int_0^1 f(\sigma_i\, n(x)z)dx = 0$ for $f \in \mathfrak{H}_0'$. Put

$$f_1(z) = \int_H k(z, z')f(z')dz' ;$$

then

$$\int_0^1 f_1(\sigma_i\, n(x)z)dx = \int_0^1 \int_H k(\sigma_i\, n(x)z, z')f(z')dz'dx$$

$$= \int_0^1 \int_H k(z, n(x)^{-1}\sigma_i^{-1}z')f(z')dz'dx$$

$$= \int_H k(z, z')\left[\int_0^1 f(\sigma_i\, n(x)z')dx\right]dz' = 0.$$

q.e.d.

It should also be noted that analytic conditions (such as the existence of integrals) were not discussed in detail in the proofs of Theorems 5.1.1 and 5.1.3 because they do not constitute an essential part of the proofs.

§5.2. Discrete and continuous spectra

The purpose of this § is to show that the space $\mathfrak{H}_0'$ defined in § 5.1 is equal to $\mathfrak{H}_0$, the closure of the space spanned by the cusp forms, and that the space Θ of incomplete theta series contains only a finite number of linearly independent eigenfunctions of D. This means that D has only a discrete spectrum in $\mathfrak{H}_0' = \mathfrak{H}_0$, and has an almost purely continuous spectrum in Θ. These results are obtained by investigating the spectrum of those integral operators (5.1.1) for which the function F is C^∞ and has compact support.

THEOREM 5.2.1. Let $K(z, z')$ be a kernel function of the type (5.1.2) such that the function F defining the point pair invariant $k(z, z')$ is C^∞ and has a compact support, let Γ be a discontinuous group of finite type, let $\kappa_1, \kappa_2, \ldots, \kappa_h$ be a complete set of inequivalent cusps lying on the boundary of a fundamental domain $\mathscr{D}$ of Γ, and let $H(z, z')$ be as in Theorem 5.1.2. Then, $K^*(z, z') = K(z, z') - H(z, z')$ is bounded on $\mathscr{D} \times \mathscr{D}$.

Proof. It is enough to prove the theorem assuming that z and z' are sufficiently close to one and the same cusp of $\mathscr{D}$. For if either (a) z is in a compact subset of $\mathscr{D}$ and z' tends to a cusp of $\mathscr{D}$, or (b) z and z' tend to different cusps of $\mathscr{D}$, then both K and H tend to zero, as one sees from formula (5.1.2) and from the formula for H in Theorem 5.1.2.

We may furthermore assume that Γ is reduced at $\infty = \kappa_1$, and may prove the theorem when z and z' both tend to ∞. In this situation,

$$K(z, z') - \sum_{\gamma \in \Gamma_0} k(z, \gamma z')$$

tends to zero. As for $H(z, z')$, we have

$$H_1(z, z') = \sum_{\sigma} \int_{-\infty}^{\infty} k(z, n(x)\, \sigma\, z')dx, \qquad (\sigma \in \Gamma_0 \backslash \Gamma),$$

and we consider a small non-euclidean circle around each point $\sigma z'$ with $0 < \operatorname{Re} \sigma z' < 1$. Then, by comparing the non-euclidean measure of a union of such circles with the non-euclidean measure of a rectangle the number $N(y_1, y_2)$ of such points satisfying $y_1 < y(\sigma z') < y_2$ may be estimated from above:

$$N(y_1, y_2) \leqq \alpha\Big(\frac{1}{y_1} - \frac{1}{y_2}\Big)$$

where α is a positive constant. Therefore, if we consider, for fixed δ,

$(0 < \delta < 1)$, a square whose vertices are z_0, $z_0 - \sqrt{-1}\delta y_0$, $z_0 - \delta y_0$, $z_0 - (\sqrt{-1} + 1)\delta y_0$, with $y_0 = \text{Im } z_0$, then the sum of the euclidean lengths of all the segments which are the intersections of the above square with the euclidean straight lines, each consisting of the complex numbers of fixed imaginary part equal to some $y(\sigma z')$, is bounded. This fact implies that $H_1(z, z')$ is bounded as z, z' tend to a common cusp κ_i different from ∞. Hence, after changing the numbering of the cusps, we may conclude that $H_i(z, z')$ with $i \neq 1$ is bounded as z, $z' \to \infty$. Since, furthermore,

$$H_1(z, z') - \int_{-\infty}^{\infty} k(z, n(x)z')dx$$

tends to zero as z and z' tend to ∞, we see finally that

$$H(z, z') - \int_{-\infty}^{\infty} k(z, n(x)z')dx$$

is bounded as z and z' tend to ∞.

Thus, it is now enough to show that

$$\sum_{\gamma \in \Gamma_0} k(z, \gamma z') - \int_{-\infty}^{\infty} k(z, n(x)z')dx$$

is bounded as z and z' tend to ∞. If $z' = x' + \sqrt{-1}y'$, then this expression is equal to

$$(5.2.1) \qquad \sum_{b \in Z} k\left(\frac{z - x'}{y'}, \sqrt{-1} + \frac{b}{y'}\right) - y' \int_{-\infty}^{\infty} k\left(\frac{z - x'}{y'}, \sqrt{-1} + t\right)dt$$

However, in general, if $f(z)$ is any C^∞ function of a real variable such that the support of f is compact and has euclidean measure M, then f has the property

$$\left|\frac{1}{y}\sum_{b \in Z} f\left(\frac{b}{y}\right) - \int_{-\infty}^{\infty} f(t)dt\right| \leq \frac{1}{y} M \max\left|\frac{d}{dt}f(t)\right|.$$

Therefore, by the assumption on $k(z, z')$, we have

$$\frac{1}{y'}\sum_{c \in Z} k\left(w, \sqrt{-1} + \frac{b}{y'}\right) - \int_{-\infty}^{\infty} k(w, \sqrt{-1} + t)dt = O\left(\frac{1}{y'}\right)$$

uniformly for $w \in H$ as $y' \to 0$. This implies that (5.2.1) is bounded as z and z' tend to ∞. *q.e.d.*

We can now prove a fundamental theorem on the space of cusp forms.

THEOREM 5.2.2. Let Γ be a discontinuous group of finite type. Let

$\mathfrak{H}_0'$ be the space consisting of all those automorphic functions with respect to Γ for which the constant terms in the Fourier expansions are zero at all cusps. Then, $\mathfrak{H}_0'$ coincides with the closed subspace $\mathfrak{H}_0$ of $\mathfrak{H} = L^2(\Gamma\backslash H)$ spanned by all cusp forms.

Proof. Denote by L the invariant integral operator defined by a point pair invariant $k(z, z')$ which is C^∞ and has a compact support with respect to z for a fixed z'. Then, we see first by Theorem 5.1.3 that $L\mathfrak{H}_0' \subset \mathfrak{H}_0'$. Theorem 5.2.1 implies, however, that the kernel K^* defines a compact operator since

$$\int_H \int_H |K^*(z, z')|^2 \, dzdz' < \infty,$$

and Theorem 5.1.2 shows that the space $L\mathfrak{H}_0'$ is equal to the image of $\mathfrak{H}_0'$ under the integral operator on $\Gamma\backslash H$ defined by K^*. The operator L is normal, because its adjoint is defined by $\overline{k(z',z)}$.† Hence, $\mathfrak{H}_0'$ is spanned by eigenfunctions of L. Considering all possible k, we obtain a commutative family of normal operators, and $\mathfrak{H}_0'$ is spanned by simultaneous eigenfunctions of them. The conclusion of this theorem now follows from Theorem 1.3.3. *q.e.d.*

The spectrum of D in Θ is continuous, except that Θ may contain a finite number of linearly independent eigenfunctions which are not cusp forms. The situation is made very clear by the following.

THEOREM 5.2.3. Let Γ be a discontinuous group of finite type, and let $\kappa_1, \kappa_2, \dots, \kappa_h$ be a complete set of inequivalent cusps of Γ. Let $\hat{\Theta}$ be the subspace of the space Θ of incomplete theta series spanned by eigenfunctions of D in Θ. Then, Θ is spanned by the residues of the Eisenstein series $E_i(z, s)$ for the cusps κ_i at the poles of $\varphi_{ii}(s)$ on the interval $\frac{1}{2} < s \leqq 1$, where $\varphi_{ii}(s)$ is the i-th diagonal element of the constant term matrix $\Phi(s)$ of the Eisenstein series

Proof. The residue $\theta_i(z)$ of $E_i(z, s)$ at a pole S_0, ($\frac{1}{2} < S_0 \leqq 1$), of $\varphi_{ii}(s)$ is an element of $\hat{\Theta}$. Since $\theta_i(z)$ is an eigenfunction of D, it belongs to $L^2(\Gamma\backslash H)$ because it is the limit of $\varphi_{ii}(s)E_i(z, s)$ as $s \to S_0$ (Theorem 4.1.4), and θ_i is orthogonal to all cusp forms by Theorem 2.3.1.

Conversely, let f be an element of Θ satisfying $Df = \lambda f$. Then, the constant term $a_{i,0}(y)$ in the Fourier expansion of $f(z)$ at κ_i must be of the form $c_i y^{1-S_0}$, ($\frac{1}{2} < S_0$), because $a_{i,0}(y)$ is a solution of the differential equation $\dfrac{d^2a}{dy^2} - \dfrac{\lambda}{y^2} a = 0$ by (2.2.12), λ must be real, and

† See remarks preceding Theorem 7.5.3.

$$\int_Y^\infty |a_{i,0}(y)|^2 \frac{dy}{y^2} < \infty$$

for a sufficiently large $Y > 0$. Assume now that $\varphi_{11}(S_0) = \dots = \varphi_{rr}(S_0) = \infty$ and that the corresponding $\theta_1, \dots, \theta_2$ are linearly independent, and assume for $r < i$ that either θ_i is a linear combination of $\theta_1, \dots, \theta_r$ or else φ_{ii} is finite. Assume, moreover, that $\det(\lim_{s\to S_0}(s - S_0)\varphi_{ij}(s)) \neq 0$, $(i, j = 1, 2, \dots, r)$; this is permissible by Theorem 2.2.1. Then, by subtracting a suitable linear combination of θ_i from f, we get an element $g(z)$ of $\hat{\Theta}$ such that $\int_0^1 g(\sigma_i z)dx = 0$ for all $i \leqq r$. For $i > r$, we have

$$\int_0^1 g(\sigma_i z)dx = c_i^* y^{1-S_0}$$

Hence, Theorem 2.3.1 implies

$$(g^Y(z), e_i^Y(z, s)) = \sum_{l=r+1}^{h} \frac{-c_l^* \varphi_{il}(s) Y^{-S_0-\bar{s}+1}}{S_0 + \bar{s} - 1} + \frac{-c_i^* Y^{-S_0+\bar{s}}}{S_0 - \bar{s}},$$

where $e_i(z,s)$ is obtained from $E_i(z, s)$ by subtracting a suitable linear combination of $E_1, \dots, E_r$, and satisfies $e_i(z, S_0) \neq \infty$.

If $c_i^* \neq 0$, then this formula gives rise to a contradiction as $s \to S_0$. Therefore, $c_i^* = 0$ for $i = r + 1, \dots, h$, i.e., $g = 0$. This implies that f is a linear combination of the functions θ_i. *q.e.d.*

From Theorem 5.1.1, Theorem 5.2.2 and Theorem 5.2.3, one now obtains

THEOREM 5.2.4. Let Γ be a discontinuous group of finite type, let Θ be the closed subspace of $\mathfrak{H} = L^2(\Gamma\backslash H)$ spanned by the incomplete theta series, and let $\mathfrak{H}_0$ be the closed subspace of $\mathfrak{H}$ spanned by the cusp forms. Furthermore, let $\hat{\Theta}$ be the finite dimensional subspace of Θ spanned by the residues of Eisenstein series at poles on the interval (½, 1], and let Θ_0 be the orthogonal complement† of $\hat{\Theta}$ in Θ. Then, $\mathfrak{H}$ is the orthogonal sum of Θ_0, $\hat{\Theta}$, and $\mathfrak{H}_0$. The spectrum of the Laplacian D is discrete in Θ and in $\mathfrak{H}_0$, and is purely continuous in Θ_0.

It may be worthy of mention here that $\hat{\Theta}$ contains the constant functions. In fact, Theorem 2.3.1 shows that a constant function is orthogonal to all cusp forms f such that $Df = \lambda f$ with $\lambda \neq 0$. But, because of the maximum principle for ordinary harmonic functions, a cusp form f with $Df = 0$ must

† For a direct characterization of this orthogonal complement Θ_0, see remark after Theorem 7.5.5.

be 0. Thus, any constant function is orthogonal to $\mathfrak{H}_0$. If the discontinuous group is the elliptic modular group, for example, then $s = 1$ is the only pole on $\frac{1}{2} < s \leqq 1$ of $\varphi_0(s)$ in (4.4.1). Therefore, the constant functions are the only eigenfunctions of D which are contained in the space Θ, i.e., the space $\hat{\Theta}$ consists of scalars.

The three spaces Θ_0, $\hat{\Theta}$, $\mathfrak{H}_0$ in Theorem 5.2.4 are admissible subspaces with respect to invariant integral operators, i.e., for any invariant integral operator defined by a point pair invariant $k(z, z')$, each of the three spaces is mapped into itself. This is obvious for $\hat{\Theta}$ and $\mathfrak{H}_0$ because of Theorem 1.3.2. As for Θ_0, the admissibility follows from the fact that Θ_0 is the orthogonal complement of $\hat{\Theta} \oplus \mathfrak{H}_0$, and from the fact that the adjoint of the integral operator defined by $k(z, z')$ is again an invariant integral operator as has been mentioned in the proof of Theorem 5.2.2.

§5.3. Continuous spectrum and Eisenstein series

In previous sections of this chapter, we discussed a kernel function $K(z, z')$ defined from a point pair invariant, and the modified kernel function $K^*(z, z') = K(z, z') - H(z, z')$, where $H(z, z')$ is in a certain sense the *noncompact part* or the *principal part* of the kernel K. Whereas this trick leads rapidly and in an elementray way to the separation of the discrete and continuous spectra of D in $\mathfrak{H} = L^2(\Gamma\backslash H)$, the function $H(z, z')$ does not have a clear, intrinsic meaning. In this §, we shall find another formula for the noncompact part of $K(z, z')$ by means of Eisenstein series, and shall obtain a quite elegant, explicit expression for the continuous spectrum of D.

First of all, we recall Theorem 1.3.2 which says that a function f on the upper half plane H is an eigenfunction of the invariant integral operator determined by any point pair invariant $k(z, z')$ whenever f is an eigenfunction of the Laplacian $D = y^2\left(\dfrac{\partial^2}{\partial x^2} + \dfrac{\partial^2}{\partial y^2}\right)$. If $Df = \lambda f$, then we have

$$\int_H k(z, z')f(z')dz' = \Lambda_k(\lambda)f(z).$$

More precisely, $\Lambda_k(\lambda)$ is a function of the complex variable λ determined uniquely by k. In fact, $k \to \Lambda_k$ is a linear transformation. We now propose to develop some explicit formulas for this linear transformation.

Denote by $u = d(z, z')$ the non-euclidean distance on H between z and z', and by $y(z)$, the imaginary part of $z \in H$. Then, since $|z - \sqrt{-1}|^2/y(z)$ is invariant under the operation of any $\sigma \in K = SO(2)$, we see that

(5.3.1) $$t(z, z') = \frac{|z - z'|^2}{y(z)\,y(z')}$$

is a point pair invariant. Moreoever, we have

$$t = e^u + e^{-u} - 2,$$

and

$$u = \log\frac{t + 2 + \sqrt{t^2 + 4t}}{2}$$

Since any point pair invariant $k(z, z')$ is a function of u, it is also a function of t. Therefore, a one-to-one correspondence between point pair invariants $k(z, z')$ and functions $k(t)$ of a positive, real variable is established by the relation

(5.3.2) $$k(z, z') = k(t(z, z')).$$

On the other hand, if we put $\lambda = s(s - 1)$, where $s = \frac{1}{2} + \sqrt{-1}r$ with a complex variable r, then

$$\lambda = -\frac{1}{4} - r^2,$$

and there is a one-to-one correspondence between the functions $\Lambda_k(\lambda)$ and even functions $h(r)$ of a complex variable, the correspondence being given by

$$\Lambda_k(\lambda) = h(r).$$

We call the linear transformation $k(t) \to h(r)$ the *Selberg transformation*, and wish to determine it in more concrete form. Of course the study of the Selberg transformation is the same thing as the study of the linear transformation $k(z, z') \to \Lambda_k(\lambda)$.

THEOREM 5.3.1. Let $k(t)$ be a function of a positive, real variable, and put

$$\int_w^\infty \frac{k(t)}{\sqrt{t - w}}\,dt = Q(w).$$

Furthermore, define $g(u)$ by $Q(w) = g(u)$ with $w = e^u + e^{-u} - 2$. Then,

(5.3.3) $$h(r) = \int_{-\infty}^{\infty} g(u)e\Big(\frac{ru}{2\pi}\Big)du, \qquad (e(x) = \exp(2\pi\sqrt{-1}x)),$$

is the Selberg transform of k. Conversely, we have

(5.3.4) $$g(u) = \frac{1}{2\pi}\int_{-\infty}^{\infty} h(r)e\Big(-\frac{ru}{2\pi}\Big)dr,$$

and

$$k(t) = -\frac{1}{\pi}\int_t^\infty \frac{dQ(w)}{\sqrt{w-t}} \tag{5.3.5}$$

Proof. We start with the formula

$$\int_H k\left(\frac{|z-z'|^2}{yy'}\right) y'^s dz' = h(r)y^s,$$

where $z = x + \sqrt{-1}y$, $z' = x' + \sqrt{-1}y'$, $y = y(z)$ and $y' = y(z')$. We have

$$\int_H k\left(\frac{|z-z'|^2}{yy'}\right) y'^s dz' = \int_0^\infty \int_{-\infty}^\infty k\left(\frac{(x-x')^2 + (y-y')^2}{yy'}\right) y'^{s-2} dx' dy'$$

$$= \int_0^\infty \int_{-\infty}^\infty k\left(\frac{y'}{y} + \frac{y}{y'} - 2 + \frac{(x-x')^2}{yy'}\right) dx' \cdot y'^{s-2} dx' dy'.$$

If we put

$$\frac{y'}{y} + \frac{y}{y'} - 2 = w, \qquad w + \frac{(x-x')^2}{yy'} = t,$$

then

$$dx' = \frac{yy'dt}{2(x'-x)},$$

and hence

$$\int_{-\infty}^\infty k\left(w + \frac{(x-x')^2}{yy'}\right) dx' = 2\int_x^\infty k\left(w + \frac{(x-x')^2}{yy'}\right) dx'$$

$$= \int_w^\infty k(t) \frac{yy'}{x-x'} dt = \sqrt{yy'} \int_w^\infty \frac{k(t)}{\sqrt{t-w}} dt.$$

Therefore, we have

$$\int_0^\infty Q(w) \sqrt{yy'} y'^{s-2} dy' = h(r) y^s.$$

Furthermore, put $y'/y = e^u$; then

$$\int_0^\infty Q(w) \sqrt{yy'} y'^{s-2} dy' = \int_{-\infty}^\infty g(u) y e^{u/2} \cdot (ye^u)^{s-2} \cdot ye^u du$$

$$= \int_{-\infty}^\infty g(u) e^{u(s-\frac{1}{2})} du \cdot y^s = \int_{-\infty}^\infty g(u) e\left(\frac{ur}{2\pi}\right) du \cdot y^s.$$

This proves (5.3.3). The formula (5.3.4) is nothing other than the inversion formula for the Fourier transformation. To prove (5.3.5), we note first that

$$-\frac{1}{\pi}\int_u^\infty \sqrt{w-u}\,dQ(w) = -\frac{1}{\pi}\int_u^\infty \sqrt{w-u}\,d\left(\int_w^\infty \frac{k(t)}{\sqrt{t-w}}dt\right)$$

$$= -\frac{1}{\pi}\left[\sqrt{w-u}\int_w^\infty \frac{k(t)}{\sqrt{t-w}}dt\right]_u^\infty + \frac{1}{\pi}\int_u^\infty\int_w^\infty \frac{1}{2\sqrt{w-u}}\frac{k(t)}{\sqrt{t-w}}dtdw$$

$$= \frac{1}{\pi}\int_u^\infty\int_u^t \frac{dw}{2\sqrt{(w-u)(t-w)}}k(t)dt = \frac{1}{2}\int_u^\infty k(t)dt.$$

Now apply $\dfrac{d}{du}$ to this equality. Then, the result on the left side is

$$-\frac{1}{\pi}\frac{d}{du}\int_u^\infty \sqrt{w-u}\,Q'(w)dw = -\frac{1}{\pi}\frac{d}{du}\int_0^\infty \sqrt{v}\,Q'(v+u)dv$$

$$= -\frac{1}{\pi}\int_0^\infty \sqrt{v}\,Q''(v+u)dv = -\frac{1}{\pi}\sqrt{v}\,Q'(v+u)\Big|_0^\infty + \frac{1}{\pi}\int_0^\infty \frac{Q'(v+u)}{2\sqrt{v}}dv$$

$$= \frac{1}{2\pi}\int_u^\infty \frac{dQ(w)}{\sqrt{w-u}},$$

while the result of differentiating the right side shows that the latter is equal to $-\frac{1}{2}k(t)$. This proves (5.3.5). *q.e.d.*

THEOREM 5.3.2. Let $k(z, z')$ be a point pair invariant, and put $n(b) = \begin{pmatrix} 1 & b \\ & 1 \end{pmatrix}$, $(b \in \boldsymbol{R})$. Then

$$\int_{-\infty}^\infty k(z, n(b)z')db = \sqrt{yy'}\,g(\log y - \log y'),$$

where $z = x + \sqrt{-1}y$, $z' = x' + \sqrt{-1}y'$, and g is as in Theorem 5.3.1.

Proof. The integral in this theorem is equal to

$$2\int_0^\infty k\left(\frac{(y-y')^2 + x^2}{yy'}\right)dx = \sqrt{yy'}\int_0^\infty \frac{1}{\sqrt{t}}k\left(\frac{(y-y')^2}{yy'} + t\right)dt,$$

if we put $\dfrac{x^2}{yy'} = t$, $dx = \dfrac{\sqrt{yy'}}{2\sqrt{t}}dt$. Putting, moreover,

$$\left(\sqrt{\frac{y}{y'}} - \sqrt{\frac{y'}{y}}\right)^2 = w = e^u + e^{-u} - 2 = (e^{u/2} - e^{-u/2})^2,$$

the above integral is seen to be equal to

$$\sqrt{yy'}\int_w^\infty \frac{k(t)}{\sqrt{t-w}}dt = \sqrt{yy'}Q(w) = \sqrt{yy'}g(u) = \sqrt{yy'}g(\log y - \log y').$$

q.e.d.

The following theorem is particularly important and interesting; it gives us an exact description of the continuous spectrum of an invariant integral operator by means of the values of Eisenstein series on the line $\operatorname{Re} s = 1/2$.

THEOREM 5.3.3 Let Γ be a discontinuous group of finite type, $\kappa_1, \kappa_2, \ldots, \kappa_h$ be a complete set of inequivalent cusps of Γ, and let $E_i(z, s)$ be the Eisenstein series for κ_i. Let $k(z, z') = k(t(z, z'))$ be a point pair invariant as in (5.3.2), and let $h(r)$ be the Selberg transform of k. Furthermore, put

$$\text{(5.3.6)} \qquad \hat{H}(z,z') = \frac{1}{4\pi}\sum_{i=1}^{h}\int_{-\infty}^{\infty} h(r)E_i(z, s)E_i(z', \bar{s})dr, \quad \left(s = \frac{1}{2} + \sqrt{-1}r\right).$$

Then, we have

$$\text{(5.3.7)} \qquad \int_{\Gamma\backslash H} \hat{H}(z, z')f(z')dz' = 0$$

for any $f \in \mathfrak{H}_0$, and $\hat{K}(z, z') = K(z, z') - \hat{H}(z, z')$, where $K(z, z') = \sum_{\sigma \in \Gamma} k(z, \sigma z')$, is bounded on $\Gamma\backslash H$, provided that k as well as h satisfies a sufficiently strong analytic condition, for example, the condition that $k(t)$ be C^∞ and have a compact support in the set of positive real numbers.

Proof. If we apply the arguments used in the proof of Theorem 3.2.1 to a cusp form $f(z)$, we see that $y(z)^a|f(\sigma_i z)|$ is decreasing as $y(z) \to \infty$ for any $a \in \boldsymbol{R}$ and for any σ_i. On the other hand, Theorem 2.3.1 implies that $(E_i^Y(z, s), f(z)) = 0$. Hence we have $(E_i(z, s), f(z)) = 0$ for any s. This proves (5.3.7).

Let us now consider the behavior of the kernel function $\hat{H}(z, z')$ as z and z' tend to a common cusp, say, to κ_1. Then, we have

$$\sum_{i=1}^{h} E_i(\sigma_1 z, s)E_i(\sigma_1 z', \bar{s})$$

$$=\sum_{i=1}^{h}(\delta_{i1}y^s + \varphi_{i1}(s)y^{1-s} + E_i^Y(\sigma_1 z, s))(\delta_{i1}y'^{\bar{s}} + \varphi_{i1}(\bar{s})y'^{1-\bar{s}} + E_i^Y(\sigma_1 z', \bar{s}))$$

with $z = x + \sqrt{-1}y$, $z' = x' + \sqrt{-1}y'$ and with $y, y' > Y$, and the fact that $\Phi(1/2 + \sqrt{-1}r)$ is unitary (Theorem 4.3.4) implies that

$$\sum_{i=1}^{h}(\delta_{i1}y^s + \varphi_{i1}(s)y^{1-s})(\delta_{i1}y'^{\bar{s}} + \varphi_{i1}(\bar{s})y'^{1-\bar{s}})$$

$$= y^{\frac{1}{2}}y'^{\frac{1}{2}}\left(\left(\frac{y}{y'}\right)^{\sqrt{-1}r} + \left(\frac{y'}{y}\right)^{\sqrt{-1}r}\right) + \varphi_{11}(\bar{s})(yy')^{\frac{1}{2}+\sqrt{-1}r} + \varphi_{11}(s)(yy')^{\frac{1}{2}-\sqrt{-1}r}.$$

The integral

$$(5.3.8) \qquad \frac{1}{4\pi}\int_{-\infty}^{\infty} h(r)y^{\frac{1}{2}}y'^{\frac{1}{2}}\left(\left(\frac{y}{y'}\right)^{\sqrt{-1}r} + \left(\frac{y'}{y}\right)^{\sqrt{-1}r}\right)dr$$

is equal to

$$\frac{1}{2\pi}\sqrt{yy'}\int_{-\infty}^{\infty} h(r)e\left(\frac{r}{2\pi}(\log y - \log y')\right)dr$$

and coincides with $\sqrt{yy'}g(\log y - \log y')$ in Theorem 5.3.2. On the other hand, it follows from a property of the Fourier transformation that the integral

$$(5.3.9) \qquad \int_{-\infty}^{\infty} h(r)\varphi_{11}(s)(yy')^{\sqrt{-1}r}dr$$

decreases rapidly as $yy' \to \infty$ so that

$$\int_{-\infty}^{\infty} h(r)(\varphi_{11}(\bar{s})(yy')^{\frac{1}{2}+\sqrt{-1}r} + \varphi_{11}(s)(yy')^{\frac{1}{2}-\sqrt{-1}r})dr$$

is bounded as $y, y' \to \infty$. Again by the arguments used in the proof of Theorem 3.2.1, the function $y^a|E_i^Y(\sigma_1 z, s)|$ is decreasing as $y \to \infty$. Hence,

$$\sum_{i=1}^{h} E_i(\sigma_1 z, s)E_i(\sigma_1 z', \bar{s}) - (\delta_{i1}y^s + \varphi_{i1}(s)y^{1-s})(\delta_{i1}y'^{\bar{s}} + \varphi_{i1}(\bar{s})y'^{1-\bar{s}})$$

tends to zero as $y, y' \to \infty$. Besides, as we have seen in the proof of Theorem 5.2.1,

$$K(z, z') - \sum_{\gamma\in\Gamma_1} k(z, \gamma z'), \ (\Gamma_i\ \{\sigma\in\Gamma\,|\,\sigma\kappa_i = \kappa_i\}),$$

is bounded as z and z' tend to κ_1, and

$$K(\sigma_1 z, \sigma_1 z') - \sum_{\gamma\in\Gamma_1} k(\sigma_1 z, \gamma\,\sigma_1 z')$$

$$= K(\sigma_1 z, \sigma_1 z') - \sum_{b\in \mathbf{Z}} k(z, n(b)z').$$

Thus, by Theorem 5.3.2, one can conclude that $\hat{K}(z, z')$ is bounded as z and z' tend to one and the same cusp.

Let us now assume that z tends to κ_i and that z' tends to κ_j with $i \neq j$, and examine the behavior of

$$(5.3.10) \qquad \sum_{l=1}^{h} E_l(\sigma_i z, s) E_l(\sigma_j z', \bar{s})$$

$$= \sum_{l=1}^{h} (\delta_{li} y^s + \varphi_{li}(s) y^{1-s} + E_l^Y(\sigma_i z, s))(\delta_{lj} y'^s + \varphi_{lj}(\bar{s}) y'^{1-\bar{s}} + E_l^Y(\sigma_i z', \bar{s}))$$

as $y, y' \to \infty$. The fact that $\Phi(\frac{1}{2} + \sqrt{-1}t)$ is unitary implies in this case that there remain no terms of the form (5.3.8), but only terms of the form (5.3.9). Thus we see that (5.3.10) is bounded as $y, y' \to \infty$. This completes the proof. *q.e.d.*

The preceding theorem means that the integral operator determined by the kernel function in (5.3.6) gives an explicit decomposition of the continuous spectrum of the integral operator determined by the kernel function $K(z, z')$. This decomposition is not of the form $\int_{-\infty}^{\infty} \lambda dE_\lambda$ with a projection E_λ as is usual in the theory of Hilbert space, but is similar to the expression of a kernel function of the Hilbert-Schmidt type by a series containing eigenfunctions and eigenvalues.

In concluding this chapter, we propose to add a rough explanation of the fact that there exists an explicit formula for the spectral decomposition for a function $f(z) \in L^2(\Gamma\backslash H)$ which is quite similar to the spectral decomposition of an invariant integral operator.

For the sake of simplicity, we assume that Γ is a discontinuous group of finite type reduced at ∞ which, we assume, is the only cusp of Γ up to equivalence. Let $\mathfrak{H} = \Theta_0 \oplus \hat{\Theta} \oplus \mathfrak{H}_0$ be the decomposition of $\mathfrak{H} = L^2(\Gamma\backslash H)$ given by Theorem 5.2.4, and assume that $\theta_\psi \in \Theta_0$. Then, under sufficiently strong analytic assumptions on the function $\psi(y)$, $(y > 0)$, we have

$$\theta_\psi(z) = \frac{1}{2\pi\sqrt{-1}} \int_{\mathrm{Re}\, s = \frac{1}{2}} L_\psi(s) E(z, s) ds,$$

as in the proof of Theorem 3.1.1, where $E(z, s)$ is the unique Eisenstein series for Γ. (See also the footnote in Theorem 5.2.4.) Now, in fashion similar to the proof of Theorem 3.1.1, we have

$$(\theta_\psi, E(z, s)) = L_\psi(s) + \varphi(1 - s) L_\psi(1 - s),$$

where $\varphi(s)$ is the function appearing in the constant term of the Fourier expansion of $E(z, s)$ at ∞. On the other hand, it follows from the functional equation of $E(z, s)$ that

$$\frac{1}{2\pi\sqrt{-1}}\int_{\text{Re } s=\frac{1}{2}} (L_\phi(s) + \varphi(1-s)L_\phi(1-s))E(z, s)ds$$

$$= \frac{1}{\pi\sqrt{-1}}\int_{\text{Re } s=\frac{1}{2}} L_\phi(s)E(z, s)ds = 2\theta_\phi(z).$$

Hence,

$$\theta_\phi(z) = \frac{1}{4\pi}\int_{-\infty}^{\infty} (\theta_\phi, E(z, s))E(z, s)dr, \quad (s = \tfrac{1}{2} + \sqrt{-1}r). \tag{5.3.11}$$

Therefore, if F_i is an orthonormal system for the space $\hat{\Theta} \oplus \mathfrak{H}_0$, then, by means of (5.3.11), we can deduce the spectral decomposition formula

$$f(z) = \sum_i (f, F_i)F_i(z) \tag{5.3.12}$$

$$+ \frac{1}{4\pi}\int_{-\infty}^{\infty} (f, E(z, s))E(z, s)dr, \quad (s = \tfrac{1}{2} + \sqrt{-1}r),$$

for a function $f(z) \in L^2(\Gamma\backslash H)$.†

This formula is convenient for studying the spectral decomposition of the Laplacian D. Namely, if F_i is an eigenfunction of D satisfying $DF_i = \lambda_i F_i$, then, at least formally, (5.3.12) yields

$$Df(z) = \sum_i \lambda_i(f, F_i)F_i(z)$$

$$+ \frac{1}{4\pi}\int_{-\infty}^{\infty} \lambda(f, E(z, s))E(z, s)dr,$$

with $\lambda = s(s-1)$.

† Many of the deeper results related to the topics of this chapter, including this formula, are contained in W. Roelke, *Das Eigenwertproblem der automorphen Formen in der hyperbolischer Ebene, I, II. Math. Ann.*, **167**, 292–337 (1966); *ibid.*, **168**, 261–324 (1967).

CHAPTER VI

Generalizations of Eisenstein series

§6.1. The function $E(g, s)_n$

In previous chapters we have seen that Eisenstein series have effective applications in the spectral theory of $L^2(\Gamma\backslash H)$, the Hilbert space of square integrable automorphic functions with respect to a discontinuous group operating on the upper half plane H. However, in order to develop a corresponding theory for $L^2(\Gamma\backslash G)$, $G = SL(2, \boldsymbol{R})$, it is necessary to make use of more general Eisenstein series than those which we have previously denoted by $E(z, s)$. As a first step, we shall define functions denoted by $E(g, s)_n$, and will study their basic properties. The spectral theory of $L^2(\Gamma\backslash G)$ turns out to be equivalent to the (partly continuous) decomposition of the regular representation of G on $L^2(\Gamma\backslash G)$ into its irreducible components.

For an element $g \in G = SL(2, \boldsymbol{R})$, we always use x, y, and θ to denote the numbers appearing in the Iwasawa decomposition $g = n(x)a(y)k(\theta)$ given in (1.1.1), while z denotes the point $x + \sqrt{-1}y$ in the upper half plane H. On the other hand, we denote by Γ a discontinuous group of finite type reduced at the point ∞ which is assumed to be the only cusp of Γ up to equivalence. In the following chapters, the main results will be proved solely for this kind of discontinuous group. The corresponding results for a more general type of discontinuous group will be stated at the end of each article, but without proof. This is because, on the one hand, restriction to the special case simplifies the description a great deal, and on the other hand, the proofs for the general case can be carried out easily as soon as one masters the techniques explained in previous chapters for treating several cusps simultaneously.

In this chapter and in the next, a *discontinuous group* Γ will not be a group of linear transformations, but will be a discrete subgroup of $G = SL(2, \boldsymbol{R})$ containing $\begin{pmatrix} -1 & \\ & -1 \end{pmatrix}$. Accordingly, notations such as $\sigma \in \Gamma_0\backslash\Gamma$, ($\Gamma_0 = \{n(b) \mid b \in \boldsymbol{Z}\}$), used frequently in previous chapters will have to be changed into $\sigma \in \pm\Gamma_0\backslash\Gamma$. (For the purpose of working with functions defined on G instead of on $H = G/K$, $(K = SO(2))$, it is convenient to change the definition of discontinuous group.)

The *Eisenstein series* $E(g, s)_n$ is now defined by

$$E(g, s)_n = \sum_{\sigma} y(\sigma z)^s e\left(\frac{n}{2\pi}(\theta + \arg(cz + d))\right), \quad (\sigma \in \pm\Gamma_0\backslash\Gamma),$$

where $y(z) = \operatorname{Im} z$, $\sigma = \begin{pmatrix} a & b \\ c & d \end{pmatrix} \in \Gamma$, $e(x)$ is as in (2.2.1), and n is an even rational integer. The convergence properties of $E(g, s)_n$ are same as those of the Eisenstein series we have dealt with in preceding chapters, i.e., it converges absolutely if and only if $\operatorname{Re} s > 1$. If $g = n(x)a(y)k(\theta)$ and $\sigma g = n(x')a(y')k(\theta')$ are Iwasawa decompositions, we have

$$z' = \frac{az + b}{cz + d}, \qquad (z = x + \sqrt{-1}y,\ z' = x' + \sqrt{-1}y'),$$

and

$$\theta' = \theta + \arg(cz+d).$$

Therefore, if we put

$$y(g, s)_n = y(z)^s e\left(\frac{n}{2\pi}\theta\right), \tag{6.1.1}$$

then

$$E(g, s)_n = \sum_{\sigma} y(\sigma g, s)_n, \qquad (\sigma \in \pm\Gamma_0\backslash\Gamma).$$

The group $G = SL(2, \boldsymbol{R})$ has a left G-invariant Riemannian metric given by the formula

$$ds^2 = \frac{1}{y^2}(dx^2 + dy^2) + \left(d\theta - \frac{dx}{2y}\right)^2,$$

and the Laplacian associated to this metric is

$$\mathscr{C} = y^2\left(\frac{\partial^2}{\partial x^2} + \frac{\partial^2}{\partial y^2}\right) + y\frac{\partial^2}{\partial x \partial \theta} + \frac{5}{4}\frac{\partial^2}{\partial \theta^2},$$

which we call in this book the *Casimir operator* of G; it is a left invariant differential operator.

The function y_n in (6.1.1) is an eigenfunction of $\mathscr{C}$ belonging to the eigenvalue $s(s - 1) - \frac{5}{4}n^2$. Consequently, $E(g, s)_n$ also satisfies

$$\mathscr{C}E(g, s)_n = \left(s(s - 1) - \frac{5}{4}n^2\right)E(g, s)_n.$$

In addition, $E(g, s)_n$ satisfies

$$\frac{\partial}{\partial \theta} E(g, s)_n = \sqrt{-1}nE(g, s)_n,$$

and this is the same thing as

$$E(gk(\theta), s)_n = E(g, s)_n e\left(\frac{n}{2\pi}\theta\right).$$

The purpose of this § is to show that $E(g, s)_n$ has properties similar to those of the Eisenstein series considered in previous chapters, i.e., properties such as having an analytic continuation and satisfying a certain functional equation. What we have to do first is to investigate the Fourier expansion of $E(g, s)_n$. With respect to the operation of Γ on $H = G/K$, the stabilizer of ∞ in Γ is exactly the group $\pm\Gamma_0$ with $\Gamma_0 = \{n(b) | b \in \boldsymbol{Z}\}$. So, there exists a Fourier expansion of the form

$$E(g, s)_n = \sum_{m=-\infty}^{\infty} a_m(g, s)_n e(mx),$$

where the coefficients a_m are given by

$$a_m(g, s)_n = \int_0^1 E(n(x)g, s)_n e(-mx)dx.$$

This expansion will be called the *Fourier expansion* at the cusp ∞ of the Eisenstein series $E(g, s)_n$. It is easy to see that $a_m(g, s)_n$ is of the form

$$a_m(g, s)_n = a_m(y, s)_n e\left(\frac{n}{2\pi}\theta\right),$$

and that $a_m(y, s)_n$ satisfies the differential equation

$$\left[\frac{d^2}{dy^2} - \left(4\pi^2m^2 + \frac{2\pi mn}{y} + \frac{s(s-1)}{y^2}\right)\right] a_m(y, s)_n = 0.$$

Therefore, $a_m(y, s)_n$ is a kind of Whittaker function, and the class of Whittaker functions is far broader than that of the Bessel functions.

We now propose to determine $a_0(y, s)_n$ explicitly. Put

(6.1.2) $$\lambda(z) = z/|z| \qquad \text{for } z \in \boldsymbol{C}, \quad z \neq 0,$$

and

$$E(z, s)_n = E(g, s)_n e\left(-\frac{n}{2\pi}\theta\right);$$

then

$$E(z, s) = \sum y(\sigma z)^s \lambda(cz + d)^n, \qquad \left(\sigma = \begin{pmatrix} a & b \\ c & d \end{pmatrix} \in \pm\Gamma_0 \backslash \Gamma\right).$$

Hence we have

$$a_0(y, s)_n = \int_0^1 E(z, s)_n dx = \int_0^1 \sum_{\sigma} \frac{y^s}{|cz + d|^{2s}} \lambda(cz + d)^n dx$$

$$= y^s + \int_{-\infty}^{\infty} \sum_{\substack{c>0 \\ d \bmod c}} \frac{y^s}{|cz + d|^{2s}} \lambda(cz + d)^n dx$$

$$= y^s + \sum_{c>0} \frac{1}{|c|^{2s}} \int_{-\infty}^{\infty} \sum_{d} \frac{y^s}{\left|z + \dfrac{d}{c}\right|^{2s}} \lambda\left(z + \frac{d}{c}\right)^n dx$$

$$= y^s + y^s \varphi_0(s) \int_{-\infty}^{\infty} \frac{\lambda(z)^n}{|z|^{2s}} dx = y^s + y^{1-s} \varphi_0(s) \int_{-\infty}^{\infty} \frac{\lambda(\sqrt{-1} + t)^n}{(1 + t^2)^s} dt,$$

where $\varphi_0(s)$ is as in (2.2.4) with $m = 0$. Furthermore,

$$\int_{-\infty}^{\infty} \frac{\lambda(\sqrt{-1} + t)^n}{(1 + t^2)^s} dt = \int_0^{\pi} \frac{(\sqrt{-1} + \cot\theta)^n}{|\operatorname{cosec}\theta|^{2s+n}} \operatorname{cosec}^2\theta \, d\theta$$

$$= 2(-1)^{n/2} \int_0^{\pi/2} \cos n\theta \cos^{2s-2}\theta \, d\theta = (-1)^{n/2} 2^{-2s+2} \frac{\pi\Gamma(2s-1)}{\Gamma\left(s + \dfrac{n}{2}\right)\Gamma\left(s - \dfrac{n}{2}\right)}$$

$$= (-1)^{n/2}\pi^{1/2} \frac{\Gamma(s)\Gamma\left(s - \dfrac{1}{2}\right)}{\Gamma\left(s + \dfrac{n}{2}\right)\Gamma\left(s - \dfrac{n}{2}\right)}.$$

Therefore, we have

$$a_0(y, s)_n = y^s + \varphi(s)_n y^{1-s} \tag{6.1.3}$$

with

$$\varphi(s)_n = (-1)^{n/2}\pi^{1/2} \frac{\Gamma(s)\Gamma\left(s - \dfrac{1}{2}\right)}{\Gamma\left(s + \dfrac{n}{2}\right)\Gamma\left(s - \dfrac{n}{2}\right)} \varphi_0(s).$$

A similar computation shows that we have

$$a_m(y, s)_n = y^{1-s}\varphi_m(s) \int_{-\infty}^{\infty} \frac{\lambda(\sqrt{-1} + t)^n}{(1 + t^2)^s} e(-mty) dt, \quad (m \neq 0). \tag{6.1.4}$$

This integral can be expressed as a Whittaker function. But, without using the theory of Whittaker functions, it can be shown that $a_m(y, s)_n$ has properties quite similar to those of $a_m(y, s) = a_m(y, s)_0$, i.e., $|a_m(y, s)|$ is rapidly decreasing when $y \to \infty$, monotonously decreasing for large y, and the integral in (6.1.4) defines an entire function on the whole s-plane. These results are derived from (6.1.4) by partial integration, as in the case of $n = 0$. (See the proof of Theorem 3.2.1.)

Let Γ be any discontinuous group. We shall call a function on $\Gamma\backslash G$ an *automorphic function*. (Previously an automorphic function meant a function on $\Gamma\backslash H$.) It is easy to see that

$$dg = \frac{1}{y^2}\, dxdyd\theta$$

is an invariant measure (Haar measure) on $G = SL(2, R)$, where x, y, and θ mean the three coordinates of g given by the Iwasawa decomposition. For two automorphic functions f_1 and f_2, the inner product can be defined by

$$(f_1, f_2) = \int_{\Gamma\backslash G} f_1(g)\overline{f_2(g)}dg,$$

and consequently, it is permissible to speak of the Hilbert space $\mathfrak{H} = L^2(\Gamma\backslash G)$ of square integrable automorphic functions.

We now come back to our special discontinuous group Γ reduced at the unique (up to equivalence) cusp ∞, and we wish to compute a certain inner product related to $E(g, s)_n$. Let $g = n(x)a(y)k(\theta)$, $(g \in G)$, be the Iwasawa decomposition, and define the function $\tilde{E}^Y(g, s)_n$ by

$$\tilde{E}^Y(g, s)_n = \begin{cases} E(g, s)_n - y(g, s)_n & \text{if} \quad y > Y, \\ E(g, s)_n & \text{if} \quad y \leqq Y. \end{cases}$$

Here we suppose that $z = x + \sqrt{-1}y$ lies in a fundamental domain $\mathscr{D}$ of Γ on H satisfying the conditions stated in Theorem 2.3.3. Then, $\tilde{E}^Y(g, s)_n$ is square integrable on $\Gamma\backslash G$ for Re $s > 1$, and

$$(\tilde{E}^Y(g, s)_n, \tilde{E}^Y(g, \bar{s}')_n) = \pi \int_{\mathscr{D}} \tilde{E}^Y(z, s)_n \overline{\tilde{E}^Y(z, \bar{s}')_n} dz$$

where

$$\tilde{E}^Y(z, s)_n = \tilde{E}^Y(g, s)_n e\left(-\frac{n\theta}{2\pi}\right).$$

Furthermore, $E(z, s)_n$ satisfies the transformation formula

$$(6.1.5) \qquad E(\sigma z, s)_n = \lambda(cz + d)^{-n} E(z, s)_n \qquad \text{for} \quad \sigma = \begin{pmatrix} a & b \\ c & d \end{pmatrix} \in \Gamma.$$

Thus, $\mathscr{D}_Y$ and $\mathscr{D}_Y'$ being as in Theorem 2.3.3, we have

$$\pi^{-1}(\tilde{E}^Y(g, s)_n, \tilde{E}^Y(g, \bar{s}')_n)$$

$$= \int_{\mathscr{D}_Y} E(z, s)_n \overline{E(z, \bar{s}')_n} dz + \int_{\mathscr{D}_Y'} (E(z, s)_n - y^s)(\overline{E(z, \bar{s}')_n} - y^{s'}) dz$$

$$= \int_{\cup \sigma \mathscr{D}_Y} y^s \overline{E(z, \bar{s}')_n} dz + \int_{\cup' \sigma \mathscr{D}_Y'} y^s \overline{E(z, \bar{s}')_n} dz + \int_{\mathscr{D}_Y'} (E(z, s)_n - y^s) y^{s'} dz,$$

where $\cup'$ means to exclude $\sigma = 1$. The domain $(\cup \sigma \mathscr{D}_Y) \cup (\cup' \sigma \mathscr{D}_Y')$, $(\sigma \in \pm \Gamma_0 \backslash \Gamma)$, is exactly the domain dctermined by $0 < x < 1$, $0 < y < Y$. Therefore, (6.1.3) yields

$$\text{(6.1.6)} \qquad \pi^{-1}(\tilde{E}^Y(g, s)_n, \tilde{E}^Y(g, \bar{s}')_n)$$

$$= \int_0^Y \int_0^1 y^s \overline{E(z, \bar{s}')_n} \frac{dxdy}{y^2} + \int_Y^\infty \int_0^1 (E(z, s)_n - y^s) y^{s'} \frac{dxdy}{y^2}$$

$$= \int_0^Y y^s (y^{s'} + \varphi(s')_n y^{1-s'}) \frac{dy}{y^2} + \int_Y^\infty \varphi(s)_n y'^{1-s} y^{s'} \frac{dy}{y^2}$$

$$= \frac{Y^{s+s'-1}}{s + s' - 1} + \frac{\varphi(s')_n Y^{s-s'}}{s - s'} - \frac{\varphi(s)_n Y^{-s+s'}}{s - s'}.$$

This calculation is valid under the assumption $\operatorname{Re} s > \operatorname{Re} s'$. Since, however, s and s' are symmetric, the above equality is valid for s, and s' with different real parts, and, after passing to the limit, (6.1.6) is valid without exception for s and s' in the region $\operatorname{Re} s > 1$.

The inner prodcut formula (6.1.6) can also be derived by using Green's formula for the Casimir operator $\mathscr{C}$. Actually, Green's formula for our inner product produces the same formula as (2.3.2). However, we already know the analytic continuation of $\varphi(s)_n$; therefore we do not need any theorem corresponding to Theorem 2.3.1. This is why we have calculated the inner product of $\tilde{E}^Y(g, s)_n$ and $\tilde{E}^Y(g, s')_n$, the computation being quite simple.

Based upon the inner product formula (6.1.6), we can repeat the whole theory of the analytic continuation given in Chap. III and Chap. IV for our function $E(g, s)_n$. Since we already have the analytic continuation of $\varphi(s)_n$, everything is obtained simply by using the idea given in Theorem 3.2.2, although some theorems of Chap. I should be used in a suitably generalized form. Such generalizations are still special cases of the results on general weakly symmetric Riemannian spaces in the sense of Selberg. As an example, we have the following: If a function on G is an eigenfunction of both $\mathscr{C}$ and $\frac{\partial}{\partial \theta}$, then it is an eigenfunction of all integral operators in G which are invariant both under the operation of G and under the transformations given by $n(x)a(y)k(\theta) \to n(x)a(y)k(\theta + \theta')$ for arbitrary θ' (*cf.* Theorem 1.3.2.).

After all the above discussion, we obtain

THEOREM 6.1.1. The Eisenstein series $E(g, s)_n$ for a discontinuous group Γ of finite type reduced at ∞, which is assumed to be the only cusp up to the equivalence, has a meromorphic continuation onto the whole s-plane, and satisfies the functional equation

$$E(g, s)_n = \varphi(s)_n E(g, 1 - s)_n,$$

where

$$\begin{aligned}\varphi(s)_n &= (-1)^{n/2}\pi^{1/2}\frac{\Gamma(s)\Gamma\left(s-\frac{1}{2}\right)}{\Gamma\left(s+\frac{n}{2}\right)\Gamma\left(s-\frac{n}{2}\right)}\varphi_0(s)\\ &= (-1)^{n/2}\frac{\Gamma(s)^2}{\Gamma\left(s+\frac{n}{2}\right)\Gamma\left(s-\frac{n}{2}\right)}\varphi(s),\end{aligned}$$

and $\varphi_0(s)$, $\varphi(s)$ are as in (2.2.7). The function $\varphi(s)_n$ satisfies $\varphi(s)_n\varphi(1-s)_n = 1$, and $|\varphi(s)_n| = 1$ if Re $s = \frac{1}{2}$. Furthermore, $E(g, s)_n$ is holomorphic for Re $s \geqq \frac{1}{2}$ except for simple poles of $\varphi(s)_n$ on the interval $(\frac{1}{2}, 1]$. At such poles, $\varphi(s)_n^{-1}E(g, s)_n$ is holomorphic.

Let Γ be a general discontinuous group of finite type, for which $\kappa_1, \kappa_2, \ldots, \kappa_h$ is a complete set of inequivalent cusps, and let Γ_i and σ_i be as in §2.1, $(i = 1, \ldots, h)$. Then, we define an Eisenstein series for each i by

(6.1.7) $$E_i(g, s)_n = \sum y(\sigma_i^{-1}\sigma g, s)_n, \qquad (\sigma \in \pm\Gamma_i\backslash\Gamma).$$

Theorem 6.1.1 can be generalized to these functions; the results take a matric form similar to that appearing in the results of §4.4. We have, among other things, the functional equation

$$\mathscr{E}(z, s)_n = \Phi(s)_n\mathscr{E}(z, 1-s)_n$$

satisfied by the column vector $\mathscr{E}(z, s) = {}^t(E_1(z, s)_n, \ldots, E_h(z, s)_n)$, where

$$\Phi(s)_n = (-1)^{n/2}\frac{\Gamma(s)^2}{\Gamma\left(s+\frac{n}{2}\right)\Gamma\left(s-\frac{n}{2}\right)}\Phi(s),$$

and $\Phi(s)$ is as in §2.2.

§6.2. The function $E\varphi(g, s)$

As before, we denote by N, A, and K respectively the subgroups of

$G = SL(2, \boldsymbol{R})$ consisting of the elements $n(x) = \begin{pmatrix} 1 & x \\ & 1 \end{pmatrix}$, $(x \in \boldsymbol{R})$, of the elements $a(y) = \begin{pmatrix} \sqrt{y} & \\ & \sqrt{y}^{-1} \end{pmatrix}$, $(y > 0)$, and of the elements $k(\theta) = \begin{pmatrix} \cos\theta & -\sin\theta \\ \sin\theta & \cos\theta \end{pmatrix}$ respectively. While G/K is identified with the upper half plane H by the natural projection $G \ni g = n(x)a(y)k(\theta) \rightarrow z = x + \sqrt{-1}y = g\sqrt{-1} \in H$, the homogeneous space $N\backslash G$ is identified with $\boldsymbol{C}^{\times} = \boldsymbol{C} - \{0\}$ by the natural projection

(6.2.1) $\quad g = n(x)a(y)k(\theta) \rightarrow y^{-\frac{1}{2}}(\cos\theta + \sqrt{-1}\sin\theta) \in \boldsymbol{C}^{\times}$

obtained by considering the second row of the matrix $a(y)k(\theta)$.

In the sequel, we denote in general by ψ a function on $\mathbf{C}^{\times} = N\backslash G$ which is invariant under $\theta \rightarrow \theta + \pi$ and tends to 0 sufficiently rapidly both as y tends to 0 and as y tends to ∞. The invariance of ψ under $\theta \rightarrow \theta + \pi$ is required in order that one may regard ψ as a function on the group $G/\{\pm 1\}$ of fractional linear transformations of H.

For the moment, we restrict ourselves to ψ of a special type such that

(6.2.2) $$\psi(g) = \psi(y)e\left(\frac{n\theta}{2\pi}\right)$$

with a function $\psi(y)$ defined for $y > 0$ and with $e\left(\frac{n\theta}{2\pi}\right) = \exp(\sqrt{-1}n\theta)$, $(n \equiv 0 \pmod 2,\ n \in \boldsymbol{Z})$. However, the definition of a new type of Eisenstein series which we now introduce remains the same for general ψ.

Put

(6.2.3) $$L_{\psi}(g, s) = \int_0^{\infty} \psi(a(y)g)y^{-s}\frac{dy}{y};$$

this will be called the *Mellin transform* of ψ as in (3.1.1). We assume that Γ is a discontinuous group of finite type reduced at ∞, and that ∞ is the only cusp of Γ up to equivalence. Denoting as before by Γ_0 the group $\{n(b) \mid b \in \boldsymbol{Z}\}$, we define the *Eisenstein series* $E_{\psi}(g, s)$ by

$$E_{\psi}(g, s) = \sum L_{\psi}(\sigma g, s), \qquad (\sigma \in \pm\Gamma_0\backslash\Gamma).$$

Put

(6.2.4) $$L_{\psi}(s) = \int_0^{\infty} \psi(y)y^{-s}\frac{dy}{y}$$

in accordance with the notation in §3.1; then, it follows from (6.2.3) that

$$L_\phi(g, s) = \int_0^\infty \varphi\left(a(y')a(y)\begin{pmatrix}1 & \\ & 1\end{pmatrix}\right)y'^{-s}\frac{dy'}{y'}\cdot e\left(\frac{n\theta}{2\pi}\right)$$
$$= L_\phi(s)y^s e\left(\frac{n\theta}{2\pi}\right) = L_\phi(s)y(g, s)_n.$$

Hence, it follows from the definition that

(6.2.5) $$E_\phi(g, s) = L_\phi(s)E(g, s)_n,$$

provided that ϕ satisfies (6.2.2). If $\phi(y)$ is rapidly decreasing both as $y \to \infty$ and as $y \to 0$, then $L_\phi(g, s)$ becomes an entire function of s. Then, the analytic continuation of $E_\phi(g, s)$ and the functional equation

(6.2.6) $$E_\phi(g, s) = L_\phi(s)\varphi(s)_n E(g, 1 - s)_n = \frac{L_\phi(s)}{L_{\phi'}(1 - s)}\varphi(s)_n E_{\phi'}(g, 1 - s)$$

for any ϕ' of type (6.2.2) are immediate consequences of the results in §6.1. A very important step in our theory is, however, to bring the above equation to a completely different and simple form which does not depend on n. For this purpose we require some analytic preliminaries related to the theory of special functions.

Let $\phi(y, \theta)$ be a function defined for $y > 0$ and $\theta \in \boldsymbol{R}$ satisfying $\phi(y, \theta + \pi) = \phi(y, \theta)$. We then define the *Fourier transform* $\hat{\phi}(y, \theta)$ of ϕ by

(6.2.7)

$$\hat{\phi}(y^{-2}, \theta) = \int_0^\infty \int_0^{2\pi} \phi(y'^{-2}, -\theta')e(\sqrt{-1}yy' \exp(\sqrt{-1}(\theta + \theta')))y'd\theta'dy'$$

with

$$e(w) = \exp(\pi\sqrt{-1}(w + \bar{w})) \qquad \text{for} \qquad w \in \boldsymbol{C}.$$

This definition looks somewhat different from the ordinary Fourier transformation. However, if y and θ in $\phi(y, \theta)$ denote the y and the θ in the Iwasawa decomposition $g = n(x)a(y)k(\theta)$ of $g \in G$, then (6.2.1) shows that the function defined by $\phi(w) = \phi(y^{-2}, \theta)$ for $w = y\exp(\sqrt{-1}\theta)$ is the function on $\boldsymbol{C}^\times = N\backslash G$ induced naturally by the function $\phi(y, \theta)$ on G. Therefore, the definition implies that

$$\hat{\phi}(w) = \int_{\boldsymbol{C}} \phi(\overline{w'})e(\sqrt{-1}ww')dV(w')$$

where $dV(w')$ is the euclidean volume $\frac{1}{2}dw' \wedge \overline{dw'}$ on $\boldsymbol{C}$. Thus, (6.2.7) is actually an ordinary Fourier transformation. Since $e(\sqrt{-1}w) = e(-\sqrt{-1}w)$, and since $\phi(-w) = \phi(w)$ by assumption, we have $\hat{\phi}(-w) = \hat{\phi}(w)$ and

$$\hat{\phi}(\bar{w}) = \int_C \phi(w')e(\sqrt{-1}ww')dV(w').$$

From this and from the inversion formula of the Fourier transformation, it follows at once that $\hat{\hat{\phi}} = \phi$.

THEOREM 6.2.1. If $\phi(y)$ is a function of a positive variable, and if $\phi(y, \theta) = \phi(y)e\left(\frac{n\theta}{2\pi}\right)$ with an even rational integer n, then the function $\hat{\phi}(y, \theta)$ given by (6.2.7) is also of the form $\hat{\phi}(y)e\left(\frac{n\theta}{2\pi}\right)$, where $\hat{\phi}(y)$ is the *Bessel transform* of order n of ϕ, i.e.,

$$\hat{\phi}(y^{-2}) = 2\pi\int_0^\infty J_n(2\pi yy')\phi(y'^{-2})y'dy', \tag{6.2.8}$$

where J_n is the standard Bessel function notation.

Proof. By (6.2.7), and from the definition of J_n, we have

$$\begin{aligned}
&\hat{\phi}(y^{-2}, \theta) \\
&= \int_0^\infty\int_0^{2\pi} \phi(y'^{-2})e\left(-\frac{n\theta'}{2\pi}\right)e(\sqrt{-1}yy' \exp(\sqrt{-1}(\theta + \theta')))y'd\theta'dy' \\
&= e\left(\frac{n\theta}{2\pi}\right)\int_0^\infty\int_0^{2\pi} \phi(y'^{-2})\exp(-\sqrt{-1}n\theta')e(\sqrt{-1}yy'\exp(\sqrt{-1}\theta'))y'd\theta'dy' \\
&= e\left(\frac{n\theta}{2\pi}\right)\int_0^\infty \phi(y'^{-2})y'\int_0^{2\pi} \exp(-\sqrt{-1}n\theta' - 2\pi\sqrt{-1}yy'\sin\theta')d\theta'dy' \\
&= e\left(\frac{n\theta}{2\pi}\right)\int_0^\infty 2\pi\phi(y'^{-2})J_n(2\pi yy')y'dy' = \hat{\phi}(y^{-2})e\left(\frac{n\theta}{2\pi}\right).
\end{aligned}$$

q.e.d.

In this theorem, as well as in the next, analytic conditions on ϕ (such as differentiability of ϕ or the rapidly decreasing nature of ϕ as $y \to 0, \infty$) are not rigorously discussed because we do not need very precise results for our applications.

THEOREM 6.2.2. If $\phi(y)$ is a function of a positive real variable y, define $L_\phi(s)$ by (6.2.4), and $\hat{\phi}$ by (6.2.8). Then, we have

$$\frac{L_\phi(s)}{L_{\hat{\phi}}(1-s)} = \pi^{-2s+1}\frac{\Gamma\left(s + \frac{n}{2}\right)}{\Gamma\left(\frac{n}{2} + 1 - s\right)}.$$

Proof. The assertion of the theorem follows from

$$L_{\hat\varphi}(1-s) = \int_0^\infty \hat\varphi(y) y^{s-1} \frac{dy}{y} = \pi \int_0^\infty \int_0^\infty J_n\Big(2\pi\sqrt{\frac{y'}{y}}\Big)\varphi(y'^{-1})dy' \cdot y^{s-1}\frac{dy}{y}$$

$$= \pi \cdot (4\pi^2)^{s-1} \int_0^\infty J_n(y^{-\frac{1}{2}}) y^{s-1}\frac{dy}{y} \int_0^\infty \varphi(y'^{-1}) y'^{s-1} dy'$$

$$= 2^{2s-1}\pi^{2s-1} \int_0^\infty J_n(y) y^{2-2s}\frac{dy}{y} \int_0^\infty \varphi(y') y'^{-s}\frac{dy'}{y'}, \qquad \Big(\frac{3}{4} < \mathrm{Re}\, s < 1\Big),$$

and from the formula

$$\int_0^\infty J_\nu(x) x^s \frac{dx}{x} = \frac{2^{s-1}\Gamma\Big(\dfrac{s+\nu}{2}\Big)}{\Gamma\Big(\dfrac{\nu - s}{2} + 1\Big)}$$

for the Mellin transformation of the Bessel function. *q.e.d.*

From (6.2.6) and from Theorem 6.2.2, we have at once

$$E_\varphi(g, s) = \frac{L_\varphi(s)}{L_{\hat\varphi}(1-s)} \varphi(s)_n E_{\hat\varphi}(g, 1-s)$$

$$= \pi^{-2s+1} \frac{\Gamma\Big(s + \dfrac{n}{2}\Big)}{\Gamma\Big(\dfrac{n}{2} + 1 - s\Big)} \cdot (-1)^{n/2} \frac{\Gamma(s)}{\Gamma\Big(s + \dfrac{n}{2}\Big)\Gamma\Big(s - \dfrac{n}{2}\Big)} \varphi(s) E_{\hat\varphi}(g, 1-s)$$

$$= \pi^{1-2s} \frac{\Gamma(s)}{\Gamma(1-s)} \varphi(s) E_{\hat\varphi}(g, 1-s).$$

This equality depends no longer on n. On the other hand, the Fourier expansion with respect to θ shows that every function φ on $C^\times = N\backslash G$ invariant under $\theta \to \theta + \pi$ is a linear combination of functions of the type (6.2.2). Thus, we obtain

THEOREM 6.2.3. Let φ be a function on $G = SL(2, \boldsymbol{R})$ depending only on the y and θ appearing in the Iwasawa decomposition $g = n(x)a(y)k(\theta)$ of $g \in G$, and invariant under $\theta \to \theta + \pi$. Assume that φ decreases sufficiently rapidly both as $y \to \infty$ and as $y \to 0$ in order that, in particular, the Mellin transform L_φ in (6.2.4) of φ be an entire function of s. On the other hand, let Γ be a discontinuous group of finite type reduced at ∞, which we suppose is the only cusp of Γ up to equivalence. Then, the Eisenstein series $E_\varphi(g, s)$ satisfies the functional equation

$$E_\phi(g, s) = \pi^{1-2s}\frac{\Gamma(s)}{\Gamma(1-s)}\varphi(s)E_{\hat{\phi}}(g, 1-s),$$

where $\varphi(s)$ is as in (2.2.7), and $\hat{\phi}$ is as in (6.2.7).

If Γ is a general discontinuous group of finite type, then we define by a formula similar to that in (6.1.7) the Eisenstein series

$$E_{i,\phi}(g, s) = \sum_\sigma L_\phi(\sigma_i^{-1}\sigma g, s), \qquad (\sigma \in \pm\Gamma_i\backslash\Gamma). \tag{6.2.9}$$

If we denote by $\mathscr{E}_\phi(g, s)$ the column vector ${}^t(E_{1,\phi}, \ldots, E_{h,\phi})$, then the functional equation in Theorem 6.2.3 holds in the generalized form

$$\mathscr{E}_\phi(g, s) = \pi^{1-2s}\frac{\Gamma(s)}{\Gamma(1-s)}\Phi(s)\mathscr{E}_{\hat{\phi}}(g, 1-s),$$

where Φ is as in §4.4.

CHAPTER VII

Spectral theory of $L^2(\Gamma\backslash G)$

§7.1. Incomplete theta series and principal series

First of all, we introduce the notion of the principal series of unitary representations of $SL(2, \boldsymbol{R})/\{\pm 1\}$. We consider even functions $\eta(w)$ of $w = r\exp(\sqrt{-1}\theta) \in \boldsymbol{C}$ such that $\eta(w) = r^{2s-2}\eta(\exp(\sqrt{-1}\theta))$, where $r > 0$ and s is a complex number. For two such functions η_1 and η_2, we define the inner product (η_1, η_2) by

$$(7.1.1) \qquad (\eta_1, \eta_2) = \int_0^{2\pi} \eta_1(\exp(\sqrt{-1}\theta))\overline{\eta_2(\exp(\sqrt{-1}\theta))}d\theta.$$

The Hilbert space of those functions η which have finite norm with respect to this inner product will be denoted by $\mathscr{H}(s)$. In virtue of the natural mapping (6.2.1), a function $\eta \in \mathscr{H}(s)$ can be viewed as a function of $g = n(x)a(y)k(\theta) \in G = SL(2, \boldsymbol{R})$, ($n(x) = \begin{pmatrix} 1 & x \\ & 1 \end{pmatrix}$, $a(y) = \begin{pmatrix} \sqrt{y} & \\ & \sqrt{y}^{-1} \end{pmatrix}$, $k(\theta) = \begin{pmatrix} \cos\theta & -\sin\theta \\ \sin\theta & \cos\theta \end{pmatrix}$, $x, \theta \in \boldsymbol{R}$, $y > 0$), satisfying

$$\eta(g) = y^{1-s}\eta(k(\theta)), \qquad \eta(k(\theta)) = \eta(k(\theta + \pi)).$$

Since this condition is equivalent to

$$\eta(n(x)a(y)g) = y^{1-s}\eta(g), \qquad \eta(-g) = \eta(g),$$

the function $T_{g_0}\eta$ given by $(T_{g_0}\eta)(g) = \eta(gg_0)$ belongs to $\mathscr{H}(s)$ for any fixed $g_0 \in G$, whenever $\eta \in \mathscr{H}(s)$. Therefore, $g \to T_g$ gives, by restriction to $\mathscr{H}(s)$, a representation of $G/\{\pm 1\}$ in the space $\mathscr{H}(s)$. If, in particular, $\operatorname{Re} s = 1/2$, then it is known from the theory of *unitary representations* that T_g becomes irreducible and unitary. This representation of $G/\{\pm 1\}$ is said to belong to the *principal series*; the representations in the principal series are parametrized by complex numbers s with $\operatorname{Re} s = 1/2$.

What we want to do next is to define an incomplete theta series for a function ϕ on $\pm N\backslash G$, and to define a canonical mapping from the space of incomplete theta series to the space $\mathscr{H}(s)$. As in Chap. VI, every detail of the theory in this chapter will be treated under the assumption that the discontinuous group is reduced at ∞, and the results for discontinuous groups of more general type will be stated at suitable places without proof. Furthermore, again as in Chap. VI, a discontinuous group should not be understood as a transformation group, but as a subgroup of $G = SL(2, \boldsymbol{R})$, and is always assumed to contain $\begin{pmatrix} -1 & \\ & -1 \end{pmatrix}$.

Let Γ be a discontinuous group of finite type reduced at ∞, the only cusp of Γ up to the equivalence. Then, the *incomplete theta series* θ_ϕ is defined by

$$\theta_\phi(g) = \sum_{\sigma \in \pm\Gamma_0\backslash\Gamma} \phi(\sigma g), \qquad (g \in G), \tag{7.1.2}$$

where ϕ is a function on $\pm N\backslash G$. We always assume that the function satisfies certain analytic conditions. The conditions will not be specified in this book, but usually it is enough to assume that ϕ has compact support on $\boldsymbol{C}^\times = N\backslash G$, or that ϕ decreases sufficiently rapidly when the variable tends to the boundary of $\boldsymbol{C}^\times$. As explained in §6.1 for Eisenstein series $E(g, s)_n$, the function $\theta_\phi(g)$ has a Fourier expansion with respect to $\Gamma_0 = \{n(b) | b \in \boldsymbol{Z}\}$; it is the Fourier expansion of θ_ϕ at ∞. In particular, the constant term is $\theta_\phi{}^0(g) = \int_0^1 \theta_\phi(n(x)g)dx$.

THEOREM 7.1.1. Let $\theta_\phi{}^0(g)$ be the constant term in the Fourier expansion at ∞ of an incomplete theta series θ_ϕ with respect to a discontinuous group Γ of finite type for which ∞ is the only cusp up to equivalence. Then, we have

$$\theta_\phi{}^0(g) = \phi(g) + \sum_{\substack{c,d \\ c \neq 0}} \frac{1}{|c|^2} \int_N \phi(\omega n(x) a(c^2) g)dx, \quad \left(\begin{pmatrix} a & b \\ c & d \end{pmatrix} \in \Gamma,\ d \bmod c\right),$$

with $\omega = \begin{pmatrix} & -1 \\ 1 & \end{pmatrix}$.

Proof. First we see that

$$\theta_\phi{}^0(g) = \int_0^1 \theta_\phi(n(x)g)dx = \int_{\Gamma_0\backslash N} \sum_{\sigma \in \pm\Gamma_0\backslash\Gamma} \phi(\sigma\, n(x)g)dx$$

$$= \int_{\Gamma_0\backslash N} \phi(n(x)g)dx + \sum_{\substack{\sigma \in \pm\Gamma_0\backslash\Gamma/\Gamma_0 \\ c \neq 0}} \int_N \phi(\sigma\, n(x)g)dx.$$

If $c > 0$, then $N\sigma N = Na(c^{-2})\omega N$, so that

$$\int_N \psi(\sigma\, n(x)g)dx = \int_N \psi(a(c^{-2})\omega n(x)g)dx$$

$$= \int_N \psi(\omega a(c^2)n(x)g)dx = \int_N \psi(\omega n(c^2x)a(c^2)g)dx$$

$$= \frac{1}{|c|^2}\int_N \psi(\omega n(x)a(c^2)g)dx.$$

Hence, we have

$$\theta_\psi{}^0(g) = \psi(g) + \sum_{\sigma\in\pm\Gamma_0\backslash\Gamma/\Gamma_0} \frac{1}{|c|^2}\int_N \psi(\omega n(x)a(c^2)g)dx, \quad \left(\sigma = \begin{pmatrix} a & b \\ c & d\end{pmatrix}\right).$$

q.e.d.

Whenever ψ, as a function on $C^\times$, decreases sufficiently rapidly as the variable tends to the boundary, θ_ψ belongs to the space $L^2(\Gamma\backslash G)$, and the Mellin transform of θ_ψ is well defined for Re $s < 0$. Instead of the Mellin transform L appearing in (6.2.3), we use a slightly modified *Mellin transform* $\hat{\theta}_\psi$ of $\theta_\psi{}^0$ defined by

(7.1.3) $$\hat{\theta}_\psi(g, s) = \int_0^\infty \theta_\psi{}^0(a(y)g)y^{s-1}\frac{dy}{y}, \qquad (\text{Re } s > 1).$$

It is easy to see that $\hat{\theta}_\psi(g, s) \in \mathscr{H}(s)$. The mapping $\theta_\psi \to \hat{\theta}_\psi$ is a standard mapping which is essential in the decomposition of regular representations.

§7.2. Inner product formulas

The purpose of this § is to prepare several explicit formulas for expressing the inner product of automorphic functions on $\Gamma\backslash G$.

First we note a simple fact about the Mellin transformation. Let $L_f(s) = \int_0^\infty f(y)y^{-s}\frac{dy}{y}$ be the Mellin transform of f, put Re $s = S$ and $s = S + \sqrt{-1}t$. Then, we have

$$L_f(S + \sqrt{-1}t) = \int_{-\infty}^\infty f(e^{y'})e^{-Sy'}e\left(\frac{-y't}{2\pi}\right)dy',$$

that is, $L_f(S + \sqrt{-1}t)$, for fixed S, is the Fourier transform of $f(e^{y'})e^{-Sy'}$. Therefore, given two functions f_1, f_2 and two real numbers S_1, S_2, the Plancherel formula for the Fourier transformation implies that

$$(7.2.1)\qquad \frac{1}{2\pi}\int_{-\infty}^{\infty} L_{f_1}(S_1 - \sqrt{-1}t)\overline{L_{f_2}(S_2 - \sqrt{-1}t)}dt$$
$$= \int_{-\infty}^{\infty} f_1(e^{y'})e^{-S_1y'} \cdot \overline{f_2(e^{y'})}e^{-S_2y'}dy' = \int_0^{\infty} f_1(y)y^{-S_1} \cdot f_2(y)y^{-S_2}\frac{dy}{y}.$$

THEOREM 7.2.1. Let θ_ϕ and $\theta_{\phi'}$ be two incomplete theta series with respect to a discontinuous group Γ of finite type reduced at ∞, the only cusp up to equivalence. Then,

$$(\theta_\phi, \theta_{\phi'}) = \frac{1}{4\pi\sqrt{-1}}\int_K\int_{\mathrm{Re}\, s=S} \hat{\theta}_\phi(k, s)\overline{L_{\phi'}(k, \bar{s})}dsd\theta, \qquad (S > 1).$$

Proof. By definition,

$$(\theta_\phi, \theta_{\phi'}) = \int_{\Gamma\backslash G} \theta_\phi(g)\overline{\theta_{\phi'}(g)}dg = \int_{\pm\Gamma_0\backslash G} \theta_\phi(g)\overline{\phi'(g)}dg$$
$$= \frac{1}{2}\int_K\int_A\int_{\Gamma_0\backslash N} \theta_\phi(nak)\phi'(nak)\frac{dxdyd\theta}{y^2} = \frac{1}{2}\int_K\int_A \theta_\phi{}^0(ak)\overline{\phi'(ak)}\frac{dxdy}{y^2}$$
$$= \frac{1}{2}\int_K\int_A \theta_\phi{}^0(ak)y^{S-1} \cdot \overline{\phi'(ak)}y^{-S}\frac{dy}{y}, \quad (n = n(x),\ a = a(y),\ k = k(\theta)).$$

Apply formula (7.2.1); then, one obtains

$$(\theta_\phi, \theta_{\phi'}) = \frac{1}{4\pi}\int_K\int_{-\infty}^{\infty} \hat{\theta}_\phi(k, s)\overline{L_{\phi'}(k, \bar{s})}dtd\theta$$
$$= \frac{1}{4\pi\sqrt{-1}}\int_K\int_{-\infty}^{\infty} \hat{\theta}_\phi(k, s)\overline{L_{\phi'}(k, \bar{s})}dsd\theta, \qquad (s = S + \sqrt{-1}t).$$

q.e.d.

Now we prove another inner product formula which expresses the inner product between an incomplete theta series and an Eisenstein series.

THEOREM 7.2.2. Let Γ be a discontinuous group of finite type reduced at ∞, the only cusp up to the eqivalence. Denote by θ_ϕ (resp. $E_{\phi'}(g, s)$) the incomplete theta series (resp. the Eisenstein series) constructed from a function ϕ(resp. ϕ') on $N\backslash G$. Then, one has

$$(\theta_\phi, E_{\phi'}(\bar{s})) = \frac{1}{2}\int_K \hat{\theta}_\phi(k, s)\overline{L_{\phi'}(k, \bar{s})}d\theta.$$

Proof. Put $n = n(x)$, $a = a(y)$, $k = k(\theta)$; then

$$(\theta_\phi, E_{\phi'}(\bar{s})) = \int_{\Gamma\backslash G} \theta_\phi(g)\overline{E_{\phi'}(g, \bar{s})}dg = \frac{1}{2}\int_{\pm\Gamma_0\backslash G} \theta_\phi(g)\overline{L_{\phi'}(g, \bar{s})}dg$$

$$= \frac{1}{2}\int_K\int_A\int_{\Gamma_0\backslash N} \theta_\phi(nak)\overline{L_{\phi'}(nak, \bar{s})}\frac{dxdyd\theta}{y^2}$$

$$= \frac{1}{2}\int_K\int_A \theta_\phi{}^0(ak)\overline{L_{\phi'}(ak, \bar{s})}\frac{dyd\theta}{y^2}$$

$$= \frac{1}{2}\int_K\int_A \theta_\phi{}^0(ak)y^{s-1}\frac{dy}{y}\overline{L_{\phi'}(k, \bar{s})}d\theta = \frac{1}{2}\int_K \hat{\theta}_\phi(k, s)\overline{L_{\phi'}(k, \bar{s})}d\theta.$$

q.e.d.

The results in these two theorems mean that the inner product in the former theorem is decomposed into a continuous sum of the inner products in the latter theorem.

§7.3. An expression for $\hat{\theta}_\phi(g, s)$ in terms of Mellin transforms

This § is devoted to the proof of a special formula (7.3.2) which expresses $\hat{\theta}_\phi(g, s)$ in (7.1.3) as a linear combination of L_ϕ and $L_{\hat{\phi}}$, where $\hat{\phi}$ is the Fourier transform introduced in § 6.2. This result is particularly important in the proof of the Plancherel formula in the next §; the result is also interesting in itself, since L_ϕ and $L_{\hat{\phi}}$ are purely geometric, and the arithmetic part of $\hat{\theta}_\phi(g, s)$ is concentrated separately in the Dirichlet series which appears in the coefficient of $L_{\hat{\phi}}$.

First we prove a theorem from which (7.3.2) is easily derived. The proof is based upon some arguments concerning special functions such as those in §6.2,

THEOREM 7.3.1. Let $\hat{\phi}$ be the Fourier transform of a function on $\pm N\backslash G$ in the sense of §6.2, where N is the group of all $n(x) = \begin{pmatrix} 1 & x \\ & 1 \end{pmatrix}$, $(x \in \boldsymbol{R})$. Then, we have

$$\int_N L_\phi(\omega n(x)g, s)dx = \pi^{-2s+(3/2)}\frac{\Gamma\left(s-\frac{1}{2}\right)}{\Gamma(1-s)}L_{\hat{\phi}}(g, 1-s),$$

where $\omega = \begin{pmatrix} & -1 \\ 1 & \end{pmatrix}$.

Proof. First we may assume that ϕ satisfies $\phi(n(x)a(y)k(\theta)) = \phi(a(y))e\left(\frac{n\theta}{2\pi}\right)$ with even n, because any ϕ is a linear combination of functions satisfying this special condition. Next, for any $n(x')$ and $a(y')$, we have

$$\int_N L_\phi(\omega n(x)n(x')a(y')g,\, s)dx = \int_N L_\phi(\omega a(y')n(xy'^{-1})g,\, s)dx$$
$$= \int_N L_\phi(a(y'^{-1})\omega n(xy'^{-1})g,\, s)dx = y'^{1-s}\int_N L_\phi(\omega n(x)g,\, s)dx,$$

and $L_{\hat\phi}(n(x')a(y')g, 1-s) = y'^{1-s}L_{\hat\phi}(g, 1-s)$. Thus, it is enough to show that

$$\text{(7.3.1)} \qquad \int_N L_\phi(\omega n(x),\, s)dx = \pi^{-2s+(3/2)}\frac{\Gamma\left(s-\frac{1}{2}\right)}{\Gamma(1-s)}L_{\hat\phi}(1,\, 1-s),$$

i.e., to prove the theorem with $g = 1$.

Now, if we put $\phi(g) = \phi(n(x)a(y)k(\theta)) = \phi(y, \theta) = \phi(y)e\left(\frac{n\theta}{2\pi}\right)$ according to (6.2.2), then the left hand side of (7.3.1) is equal to

$$\int_{-\infty}^{\infty}\int_0^{\infty}\phi\left(\frac{y}{|\sqrt{-1}+x|^2},\quad \arg(\sqrt{-1}+x)\right)y^{-s}\frac{dy}{y}dx$$
$$= \int_{-\infty}^{\infty}\int_0^{\infty}\phi\left(\frac{y}{1+x^2}\right)e\left(\frac{n}{2\pi}\arg(\sqrt{-1}+x)\right)y^{-s}\frac{dy}{y}dx$$
$$= L_\phi(s)\int_{-\infty}^{\infty}\frac{\lambda(\sqrt{-1}+x)^n}{(1+x^2)^s}dx,$$

where λ is as in (6.1.2), and $L_\phi(s)$ is as in (6.2.4). Therefore, as in the proof of (6.1.3), we obtain

$$\int_{-\infty}^{\infty} L_\phi(\omega n(x),\, s)dx = (-1)^{n/2}\pi^{1/2}\frac{\Gamma(s)\Gamma\left(s-\frac{1}{2}\right)}{\Gamma\left(s+\frac{n}{2}\right)\Gamma\left(s-\frac{n}{2}\right)}L_\phi(s).$$

On the other hand, it follows from Theorem 6.2.2 that

$$L_\phi(s) = \pi^{-2s+1}\frac{\Gamma\left(s+\frac{n}{2}\right)}{\Gamma\left(\frac{n}{2}+1-s\right)}L_{\hat\phi}(1-s).$$

Therefore, we have

$$(-1)^{n/2}\pi^{1/2}\frac{\Gamma(s)\Gamma\left(s-\frac{1}{2}\right)}{\Gamma\left(s+\frac{n}{2}\right)\Gamma\left(s-\frac{n}{2}\right)}L_\varphi(s)$$

$$=(-1)^{n/2}\pi^{-2s+(3/2)}\frac{\Gamma(s)\Gamma\left(s-\frac{1}{2}\right)}{\Gamma\left(s-\frac{n}{2}\right)\Gamma\left(1-\left(s-\frac{n}{2}\right)\right)}L_{\hat{\varphi}}(1-s)$$

$$=\pi^{-2s+(3/2)}\frac{\Gamma(s)\Gamma\left(s-\frac{1}{2}\right)}{\Gamma(s)\Gamma(1-s)}L_{\hat{\varphi}}(1-s)=\pi^{-2s+(3/2)}\frac{\Gamma\left(s-\frac{1}{2}\right)}{\Gamma(1-s)}L_{\hat{\varphi}}(1,1-s).$$

q.e.d.

THEOREM 7.3.2. Let Γ be a discontinuous group of finite type reduced at ∞, the only cusp of Γ up to the equivalence, and denote by $\varphi(s)$ the function in (2.2.7). Then, $\hat{\theta}_\varphi(g, s)$ is a linear combination of $L_\varphi(g, s)$ and $L_{\hat{\varphi}}(g, s)$ as expressed by the following formula:

$$\hat{\theta}_\varphi(g,s)=L_\varphi(g,1-s)+\pi^{1-2s}\frac{\Gamma(s)}{\Gamma(1-s)}\varphi(s)L_{\hat{\varphi}}(g,1-s) \tag{7.3.2}$$

Proof. From Theorem 7.1.1, it follows that

$$\hat{\theta}_\varphi(g,s)=L_\varphi(g,1-s)+\sum_{c,d}\frac{1}{|c|^2}\int_0^\infty\int_N\varphi(\omega n(x)a(c^2)a(y)g)dx\,y^{s-1}\frac{dy}{y}.$$

Now, Theorem 7.3.1 implies that

$$\int_0^\infty\int_N\varphi(\omega n(x)a(c^2)a(y)g)y^{s-1}dx\frac{dy}{y}=|c|^{2-2s}\int_N L_\varphi(\omega n(x)g,s)dx$$

$$=|c|^{2-2s}\pi^{-2s+(3/2)}\frac{\Gamma\left(s-\frac{1}{2}\right)}{\Gamma(s-1)}L_{\hat{\varphi}}(g,1-s).$$

Hence,

$$\hat{\theta}_\varphi(g,s)=L_\varphi(g,1-s)+\pi^{1/2}\frac{\Gamma\left(s-\frac{1}{2}\right)}{\Gamma(s)}\sum_{c,d}\frac{1}{|c|^{2s}}\cdot\pi^{1-2s}\frac{\Gamma(s)}{\Gamma(1-s)}L_{\hat{\varphi}}(g,1-s).$$

q.e.d.

This theorem furnishes a good estimation of $\hat{\theta}_\varphi(g, s)$ in the region $\mathrm{Re}\, s \geqq 1/2$. A theorem on $\varphi(s)$ is necessary to describe the situation precisely.

THEOREM 7.3.3. Let $\varphi(s)$ be the function in (2.2.7) defined for a discontinuous group Γ of finite type reduced at ∞, the only cusp of Γ up to the equivalence. Then, $\varphi(s)$ is bounded in the region determined by $\frac{1}{2} \leqq \operatorname{Re} s < c$, $|\operatorname{Im} s| > c'$, where c and c' are arbitrary real constants such that $\frac{1}{2} < c$, $c' > 0$.

Proof. By Theorem 2.3.2,

$$\frac{Y^{2S-1} - |\varphi(s)|^2 Y^{-2S+1}}{2S-1} + \frac{\varphi(\bar{s}) Y^{2\sqrt{-1}t} - \varphi(\bar{s}) Y^{-2\sqrt{-1}t}}{2\sqrt{-1}t},$$

$$\left(s = S + \sqrt{-1}t, \quad S > \frac{1}{2}\right),$$

is the norm of a certain automorphic function, and must therefore be positive. Hence, putting $\varphi(s) = |\varphi(s)|\exp(\sqrt{-1}\alpha)$, we have

$$\frac{Y^{2S-1} - |\varphi(s)|^2 Y^{-2S+1}}{2S-1} + |\varphi(s)| \frac{\sin(-\alpha + 2t \log Y)}{t} > 0.$$

This implies that

$$-Y^{-2S+1}|\varphi(s)|^2 + (2S-1)\frac{\sin(-\alpha + 2t \log Y)}{t} |\varphi(s)| + Y^{2S-1} > 0.$$

Since the coefficients of this quadratic inequality are bounded if s is as stated in the theorem, $|\varphi(s)|$ must also be bounded. *q.e.d.*

Under the assumption that every function ϕ in L_ϕ or in θ_ϕ decreases sufficiently rapidly when the variable tends to the boundary of $\boldsymbol{C}^\times = N\backslash G$, the Mellin transforms L_ϕ and $L_{\hat{\phi}}$ are entire functions of s which decrease rapidly when Im s tends to ∞ while Re s remains in a compact interval. On the other hand, it is known that

$$|\Gamma(s)| = \sqrt{2\pi} e^{-\pi t/2} t^{S-(1/2)} (1 + O(t^{-1}))$$

holds uniformly when $\operatorname{Re} s = S$ is bounded. Therefore, by virtue of (7.3.2) $\hat{\theta}_\phi(g, s)$ is uniformly bounded in the domain given by $\frac{1}{2} \leqq \operatorname{Re} s \leqq c$, $|\operatorname{Im} s| > c'$ with any $c > \frac{1}{2}$ and $c' > 0$. This is one of the most important consequences of (7.3.2), and will be used quite effectively in the proof of the Plancherel formula in the next §.

§7.4. The Plancherel formula

Except for some discussion of generalizations which occurs after Theorem 7.4.2, we assume in this § that our discontinuous group is of finite type and reduced at ∞, which is assumed to be the only cusp of Γ up to the equivalence. Every function ϕ on $\pm N\backslash G$ appearing in this § will be assumed to be rapidly decreasing when the variable tends to the boundary of $\boldsymbol{C}^\times = N\backslash G$.

THEOREM 7.4.1. The Eisenstein series $E_\phi(g, s)$ is holomorphic for $\operatorname{Re} s \geqq 1/2$ if and only if $\theta_\phi(g, s)$ is holomorphic in the same region.

Proof. Assume ϕ to be of the type $\phi(g) = \phi(y)e\left(\frac{n\theta}{2\pi}\right)$ for $g = n(x)a(y)k(\theta)$. Then, by Theorem 6.1.1. and by (6.2.5), $E_\phi(g, s)$ is holomorphic for $\operatorname{Re} s \geqq 1/2$ if and only if $L_\phi(s)\varphi(s)_n$ is holomorphic for the same s, where $\varphi(s)_n$ is as in (6.1.3). On the other hand, we already have an explicit formula in Theorem 7.3.2 for $\hat{\theta}_\phi(g, s)$, and the proof of Theorem 7.3.1 contains a proof that

$$\pi^{1-2s}\frac{\Gamma(s)}{\Gamma(1-s)}\varphi(s)L_{\hat{\phi}}(1, 1-s) = \frac{\Gamma(s)}{\Gamma(1-s)}\frac{\Gamma\left(\frac{n}{2}+1-s\right)}{\Gamma\left(s+\frac{n}{2}\right)}\varphi(s)L_\phi(1, s)$$

$$= (-1)^{n/2}\frac{\Gamma(s)^2}{\Gamma\left(s+\frac{n}{2}\right)\Gamma\left(s-\frac{n}{2}\right)}\varphi(s)L_\phi(1, s) = \varphi(s)_n L_\phi(1, s).$$

Since in general ϕ is a linear combination of functions ϕ of the special type considered above, the theorem is proved. *q.e.d.*

The closed subspace of $\mathfrak{H} = L^2(\Gamma\backslash G)$ spanned by all θ_ϕ for which ϕ satisfies the condition of Theorem 7.4.1 will be denoted bt Θ_0, while the closed subspace spanned by all incomplete theta series is denoted by Θ. Then, the Plancherel formula for the projection $\Theta_0 \ni \theta_\phi \rightarrow \hat{\theta}_\phi(g, s) \in \mathscr{H}(s)$ is the following:

THEOREM 7.4.2. Let θ_ϕ be an incomplete theta series for which the function ϕ on $\boldsymbol{C}^\times = N\backslash G$ has the property that $E_\phi(g, s)$ has no pole on the interval $1/2 \leqq s < 1$. Then,

$$||\theta_\phi||^2 = \frac{1}{4\pi\sqrt{-1}}\int_{\substack{\operatorname{Im} s>0\\ \operatorname{Re} s=1/2}} ||\hat{\theta}_\phi||^2 ds,$$

where $||\hat{\theta}_\phi||$ denotes the norm of $\hat{\theta}_\phi(g, s)$ in $\mathscr{H}(s)$.

Proof. Theorem 7.2.1 yields

$$4\pi\sqrt{-1}||\theta_\phi||^2 = \int_K\int_{\operatorname{Re} s=S>1}\hat{\theta}_\phi(k, s)\overline{L_\phi(k, \bar{s})}ds d\theta,$$

and, by Theorem 7.3.2, Theorem 7.3.3, and Theorem 7.4.1, we may shift the intergral on Re $s = S > 1$ to the integral on Re $s = \frac{1}{2}$. Thus, it follows from Theorem 6.2.3, Theorem 7.2.2, and Theorem 7.3.2 that

$$4\pi\sqrt{-1}||\theta_\varphi||^2 = \int_K\int_{\operatorname{Re} s=\frac{1}{2}} \hat{\theta}_\varphi(k,s)\overline{L_\varphi(k,\bar{s})}dsd\theta$$

$$= 2\int_{\operatorname{Im} s>0}(\theta_\varphi, E_\varphi(\bar{s}))ds + 2\int_{\operatorname{Im} s<0}(\theta, E_\varphi(\bar{s}))ds$$

$$= 2\int_{\operatorname{Im} s>0}(\theta_\varphi, E_\varphi(g,\bar{s}))ds + 2\int_{\operatorname{Im} s>0}\left(\theta_\varphi, \pi^{1-2s}\frac{\Gamma(s)}{\Gamma(1-s)}\varphi(s)E_{\hat{\varphi}}(g,\bar{s})\right)ds$$

$$= \int_K\int_{\operatorname{Im} s>0}\hat{\theta}_\varphi(k,s)\operatorname{conj}\left[L_\varphi(k,1-s)+\pi^{1-2s}\frac{\Gamma(s)}{\Gamma(1-s)}\varphi(s)L_{\hat{\varphi}}(k,1-s)\right]dsd\theta$$

$$= \int_K\int_{\operatorname{Im} s>1}\hat{\theta}_\varphi(k,s)\overline{\hat{\theta}_\varphi(k,s)}dsd\theta = \int_{\substack{\operatorname{Im} s>0\\ \operatorname{Re} s=1/2}}||\hat{\theta}_\varphi||^2ds,$$

where "conj" denotes complex conjugation. *q.e.d.*

Let us note here the form taken by the Plancherel formula for a general discontinuous group of finite type. We denote by Γ ($\ni \pm 1$) a discontinuous group of finite type, by $\kappa_1, \ldots, \kappa_h$ a complete set of inequivalent cusps, and let Γ_i and σ_i be as in §2.1. Instead of the function θ_φ appearing in (7.1.2), we consider in this general case the incomplete theta series

(7.4.1)
$$\theta_{i,\varphi}(g) = \sum_{\sigma\in\pm\Gamma_i\backslash\Gamma}\psi(\sigma_i^{-1}\sigma g),$$

where ψ is as in (7.1.2). Correspondingly we put

$$\theta_{ij,\varphi}{}^0(g) = \int_0^1\theta_{i,\varphi}(\sigma_j n(x)g)dx,$$

and

(7.4.2)
$$\hat{\theta}_{ij,\varphi}(g,s) = \int_0^\infty\theta_{ij,\varphi}{}^0(a(y)g)y^{s-1}\frac{dy}{y}.$$

Then, Theorem 7.2.1 yields

$$(\theta_{i,\varphi}, \theta_{j,\varphi'}) = \frac{1}{4\pi\sqrt{-1}}\int_K\int_{\operatorname{Re} s=S>1}\hat{\theta}_{ij,\varphi}(k,s)\overline{L_{\varphi'}(k,\bar{s})}dsd\theta,$$

and Theorem 7.2.2 becomes

$$(\theta_{i,\varphi}, E_{j,\varphi'}(\bar{s})) = \frac{1}{2}\int_K\hat{\theta}_{ij,\varphi}(k,s)\overline{L_{\varphi'}(k,\bar{s})}d\theta$$

with $E_{j,\varphi'}$ defined by (6.2.9). Furthermore, Theorem 7.1.1 yields

$$\theta_{ij,\varphi}{}^0(g) = \delta_{ij}\psi(g) + \sum_{\sigma\in\pm\Gamma_i\backslash\Gamma/\Gamma_j} \frac{1}{|c|^2}\int_{-\infty}^{\infty} \psi(\omega n(x)a(c^2)g)dx, \quad \left(\sigma = \begin{pmatrix} a & b \\ c & d \end{pmatrix}\right),$$

and consequently Theorem 7.3.2 becomes

$$\hat{\theta}_{ij,\varphi}(g, s) = \delta_{ij}L_\varphi(g, 1-s) + \pi^{1-2s}\frac{\Gamma(s)}{\Gamma(1-s)}\varphi_{ij}(s)L_{\hat{\varphi}}(g, 1-s)$$

with $(\varphi_{ij}(s)) = \Phi(s)$ as in §4.4. From these results and from the generalization of Theorem 6.2.3 at the end of Chap. VI, it follows that

$$4\pi\sqrt{-1}(\theta_{i,\varphi}, \theta_{j,\varphi'}) = \int_K \int_{\mathrm{Re}\, s=1/2} \hat{\theta}_{ij,\varphi}(k, s)\overline{L_{\varphi'}(k, \bar{s})}dsd\theta$$

$$= 2\int_{\mathrm{Im}\, s>0} (\theta_{i,\varphi}, E_{j,\varphi'}(g, \bar{s}))ds + 2\int_{\mathrm{Im}\, s<0} (\theta_{i,\varphi}, E_{j,\varphi'}(\bar{s}))ds$$

$$= 2\int_{\mathrm{Im}\, s>0} (\theta_{i,\varphi}, E_{j,\varphi'}(g, \bar{s}))ds + 2\int_{\mathrm{Im}\, s>0} (\theta_{i,\varphi}, \pi^{1-2s}\frac{\Gamma(s)}{\Gamma(1-s)} \times \sum_l \varphi_{jl}(s)E_{l,\hat{\varphi}'}(g, \bar{s}))ds$$

$$= \int_K \int_{\mathrm{Im}\, s>0} \hat{\theta}_{ij,\varphi}(k, s)\overline{L_{\varphi'}(k, 1-s)} + \sum_l \hat{\theta}_{il,\varphi}(k, s)\mathrm{conj}\left[\pi^{1-2s}\frac{\Gamma(s)}{\Gamma(s-1)}\varphi_{jl}(s)L_{\hat{\varphi}'}(k, 1-s)\right]dsd\theta$$

$$= \sum_l \int_K \int_{\mathrm{Im}\, s>0} \hat{\theta}_{il,\varphi}(k, s)\mathrm{conj}\left[\delta_{jl}L_{\varphi'}(k, 1-s) + \pi^{1-2s}\frac{\Gamma(s)}{\Gamma(1-s)} \times \varphi_{jl}(s)L_{\hat{\varphi}'}(k, 1-s)\right]dsd\theta$$

$$= \sum_l \int_{\mathrm{Im}\, s>0} \int_K \hat{\theta}_{il,\varphi}(k, s)\overline{\hat{\theta}_{jl,\varphi'}(k, s)}dsd\theta,$$

provided that $\hat{\theta}_{ij,\varphi}(g, s)$ is holomorphic on $\mathrm{Re}\, s \geqq 1/2$. Thus, in particular, we obtain

$$||\theta_{i,\varphi}||^2 = \frac{1}{4\pi\sqrt{-1}}\sum_{l=1}^{h}\int_{\substack{\mathrm{Re}\, s=1/2 \\ \mathrm{Im}\, s>0}} ||\hat{\theta}_{il,\varphi}||^2 ds, \tag{7.4.3}$$

which is the Plancherel formula for the general case. This is so to speak an h-dimensional spectral decomposition.

For the sake of completeness in the theory, it is desirable to give a theoretical characterization of the image consisting of the functions $\hat{\theta}_\phi(g, s) \in \mathscr{X}(s)$ in the Plancherel formula, namely to determine what functions of two variables g, s can be of the form $\hat{\theta}_\phi(g, s)$. In this book, no detailed investigation of this problem will be given, but we add here a rough derivation of an inversion formula for the mapping $\theta_\phi \to \hat{\theta}_\phi(g, s)$, which can be regarded as partial information on the above mentioned image space.

Because of the formula for the inverse Mellin transformation, (6.2.3) is equivalent to

$$\phi(a(y)g) = \frac{1}{2\pi\sqrt{-1}} \int_{\operatorname{Re} s = S} L_\phi(g, s) y^s ds.$$

So, in particular,

$$\phi(g) = \frac{1}{2\pi\sqrt{-1}} \int_{\operatorname{Re} s = S} L_\phi(g, s) ds,$$

and

$$\theta_\phi(g) = \frac{1}{2\pi\sqrt{-1}} \int_{\operatorname{Re} s = S > 1} E_\phi(g, s) ds. \tag{7.4.4}$$

If $\theta_\phi \in \Theta_0$, then this integral can be shifted to the integral on $\operatorname{Re} s = \frac{1}{2}$, and by the functional equation of E, one obtains

$$\theta_\phi(g) = \frac{1}{4\pi\sqrt{-1}} \int_{\operatorname{Re} s = 1/2} E_\phi(g, 1-s) + \pi^{1-2s} \frac{\Gamma(s)}{\Gamma(1-s)} \varphi(s) E_{\hat{\phi}}(g, 1-s) ds.$$

However, by Theorem 7.3.2,

$$E_\phi(g, 1-s) + \pi^{1-2s} \frac{\Gamma(s)}{\Gamma(1-s)} \varphi(s) E_{\hat{\phi}}(g, 1-s) = \sum_{\sigma \in \pm\Gamma_0\backslash\Gamma} \hat{\theta}_\phi(\sigma g, s).$$

Therefore, the mapping inverse to $\theta_\phi(g) \to \hat{\theta}_\phi(g, s)$ is given by

$$\theta_\phi(g) = \frac{1}{4\pi\sqrt{-1}} \int_{\operatorname{Re} s = 1/2} \sum_{\sigma \in \pm\Gamma_0\backslash\Gamma} \hat{\theta}_\phi(\sigma g, s) ds.$$

§7.5. Spectral decomposition of regular representations

Except for some comments at the end, we assume throughout this § that our discontinuous group is of finite type, contains ± 1, and is reduced at ∞ which is assumed to be the only cusp up to the equivalence.

Let $\mathfrak{H} = L^2(\Gamma\backslash G)$ be the space of square integrable automorphic functions on $G = SL(2, R)$ with respect to Γ. Then, $T_{g_0} : f(g) \to f(gg_0)$ is a unitary representation of G in $\mathfrak{H}$, which we shall call the *regular representation* of G by $\mathfrak{H}$. In §7.4, we defined a particular closed subspace Θ_0 of $\mathfrak{H}$; the Plancherel formula (Theorem 7.4.2) shows that the regular representation restricted on Θ_0 is a continuous sum of irreducible representations in the principal series. But, the space Θ_0 does not coincide with $\mathfrak{H}$. As a matter of fact, T_g also has a discrete spectrum in $\mathfrak{H}$. To clarify the situation, we now introduce two other closed subspaces $\hat{\Theta}$ and $\mathfrak{H}_0$ of $\mathfrak{H}$, such that Θ, the space of incomplete theta series, is the orthogonal sum of Θ_0 and of $\hat{\Theta}$, and $\mathfrak{H}$ is the orthogonal sum of Θ and of $\mathfrak{H}_0$. The space $\hat{\Theta}$, as well as $\mathfrak{H}_0$, will be decomposed into the discrete sum of irreducible representation spaces. The situation is quite similar to the spectral decomposition of invariant integral operators on the upper half plane.

The space $\mathfrak{H}_0$ is defined as the space of all automorphic functions f in $\mathfrak{H}$ such that

$$\int_0^1 f(n(x)g)dx = 0.$$

A function $f \in \mathfrak{H}_0$ will be called a *cusp form* when f is an eigenfunction of the Casimir operator $\mathscr{C}$ and of the operator $\dfrac{\partial}{\partial\theta}$. This means that, in particular, a cusp form is an eigenfunction of $\mathscr{C}$ satisfying $f(gk(\theta)) = f(g)e\left(\dfrac{n\theta}{2\pi}\right)$ for some even $n \in \boldsymbol{Z}$, where $k(\theta)=\begin{pmatrix}\cos\theta & -\sin\theta\\ \sin\theta & \cos\theta\end{pmatrix}$ and $e(x)=\exp(2\pi\sqrt{-1}x)$ as before.

The space of all $f \in \mathfrak{H}_0$ satisfying $f(gk(\theta)) = f(g)e\left(\dfrac{n\theta}{2\pi}\right)$ for some even n will be denoted by $\mathfrak{H}_{0,n}$.

The following theorem gives a natural identification of the cusp forms with certain functions on the upper half plane H which are usually called *automorphic forms*.

THEOREM 7.5.1. Let f be a cusp form belonging to $\mathfrak{H}_{0,n}$, and let $g = n(x)a(y)k(\theta)$, $\left(n(x) = \begin{pmatrix}1 & x\\ & 1\end{pmatrix}, a(y) = \begin{pmatrix}\sqrt{y} & \\ & \sqrt{y}^{-1}\end{pmatrix}, k(\theta) = \begin{pmatrix}\cos\theta & -\sin\theta\\ \sin\theta & \cos\theta\end{pmatrix}\right)$, be the Iwasawa decomposition. Then, the function $f_H(z) = y^{n/2}f(g)e\left(-\dfrac{n\theta}{2\pi}\right)$ is a function on the upper half plane H satisfying $\displaystyle\int_0^1 f_H(n(x)z)dx = 0$ and $f_H(\sigma z) = (cz+d)^{-n}f_H(z)$ for $\sigma = \begin{pmatrix}a & b\\ c & d\end{pmatrix} \in \Gamma$, $(z \in H)$. Moreover, f_H is rapidly decreasing as $y \to \infty$ and, if $\mathscr{C}f = \lambda_{\mathscr{C}}f$, where $\mathscr{C}$ is the Casimir operator, then

$$Df_H + \sqrt{-1}ny\left(\frac{\partial}{\partial x} + \sqrt{-1}\frac{\partial}{\partial y}\right)f_H = \lambda_H f_H$$

with D, the Laplacian and with $\lambda_H = \lambda_{\mathscr{C}} + n^2 - \frac{n}{2}$. Conversely, for every f_H satisfying all these conditions, $f(g) = y^{-n/2}f_H(z)e\left(\frac{n\theta}{2\pi}\right)$ is a cusp form in $\mathfrak{H}_{0,n}$.

Proof. The transformation formula for f_H is virtually equivalent to (6.1.5). If $\mathscr{C}f = \lambda_{\mathscr{C}}f$, then

$$\left(D + y\frac{\partial^2}{\partial x\partial\theta} + \frac{5}{4}\frac{\partial^2}{\partial\theta^2}\right)y^{-n/2}f_H(z)e\left(\frac{n\theta}{2\pi}\right)$$

$$= y^{-n/2}\left(y^2\frac{\partial^2}{\partial x^2} + \frac{n}{2}\left(\frac{n}{2}+1\right) - ny\frac{\partial}{\partial y} + y^2\frac{\partial^2}{\partial y^2} + \sqrt{-1}ny\frac{\partial}{\partial x} - \frac{5}{4}n^2\right) \times f_H(z)e\left(\frac{n\theta}{2\pi}\right)$$

$$= \lambda_{\mathscr{C}}y^{-n/2}f_H(z)e\left(\frac{n\theta}{2\pi}\right).$$

It follows from this that

$$\left(D + ny\left(-\frac{\partial}{\partial y} + \sqrt{-1}\frac{\partial}{\partial x}\right) - n^2 + \frac{n}{2}\right)f_H = \lambda_{\mathscr{C}}f_H\,.$$

To prove the other assertions, see the proof of Theorem 3.2.1. (Whittaker functions will appear instead of the function K.) *q.e.d.*

THEOREM 7.5.2. The space $\mathfrak{H}_0$ is the orthogonal complement of the space Θ of incomplete theta series in $\mathfrak{H}$.

Proof. Let $\theta_\varphi = \sum_{\sigma\in\pm\Gamma\backslash G}\varphi(\sigma g)$, and let f be an arbitrary function in $\mathfrak{H} = L^2(\Gamma\backslash G)$. Then,

$$(\theta_\varphi, f) = \int_{\Gamma\backslash G}\theta_\varphi(g)f(g)dg = \frac{1}{2}\int_{\Gamma_0\backslash G}\varphi(g)f(g)dg$$

$$= \frac{1}{2}\int_{\Gamma_0\backslash N}\int_A\int_K\varphi(nak)f(nak)dg = \frac{1}{2}\int_A\int_K\varphi(ak)f^0(ak)\frac{dyd\theta}{y^2}$$

with $f^0(g) = \int_0^1 f(n(x)g)dx$. Thus, the condition that $(\theta_\varphi, f) = 0$ for all φ is equivalent to $f^0 = 0$. *q.e.d.*

Denote by F a C^∞ function with compact support on G such that $F(g) = F(-g)$. Such a function defines the kernel function K of an integral operator T_F on $\mathfrak{H} = L^2(\Gamma\backslash G)$ if we put

$$K(g, g') = \sum_{\sigma \in \Gamma} k(g, \sigma g'), \qquad \text{with } k(g, g') = F(g'^{-1} g),$$

and

$$(T_F f)(g) = \int_{\Gamma\backslash G} K(g, g') f(g') dg', \qquad (f \in \mathfrak{H}).$$

Since

$$(T_F f)(g) = \int_G F(g'^{-1}g) f(g') dg' = \int_G f(gg') F(g'^{-1}) dg',$$

$T_F f$ is the convolution $f*F$ of f and F on G, and, using the operator $T_{g'}$ defined at the beginning of this §, we have

$$T_F = \int_G F(g'^{-1}) T_{g'} dg' ,$$

that is, T_F is a continuous sum of unitary operators $T_{g'}$. For many problems, the investigation of $T_{g'}$ can be replaced by that of T_F.

Put now

$$H(g, g') = \sum_{\sigma \in \Gamma_0\backslash\Gamma} \int_{-\infty}^{\infty} k(g, n(x)\sigma g') dx. \tag{7.5.1}$$

Then, for any $f \in \mathfrak{H}_0$, we have

$$\int_{\Gamma\backslash G} H(g, g') f(g') dg' = \int_{\Gamma\backslash G} \sum_{\sigma \in \Gamma_0\backslash\Gamma} \int_N k(g, n\sigma g') dx f(g') dg'$$

$$= \int_{\Gamma_0\backslash G} \int_N k(g, ng') f(g') dx dg'$$

$$= \int_{N\backslash G} \int_{\Gamma_0\backslash N} \int_N k(g, nn'a'k') f(n'a'k') dx dx' \frac{dy' d\theta'}{y'^2}$$

$$= \int_{N\backslash G} \left[\int_N k(g, na'k') dx \cdot \int_{\Gamma_0\backslash N} f(n'a'k') dx' \right] \frac{dy' d\theta'}{y'^2} = 0.$$

Hence, there prevails a situation quite similar to that encountered in Theorem 5.1.2, and indeed arguments similar to those used in Theorem 5.2.1 lead to the conclusion that $K(g, g') - H(g, g') = K^*(g, g')$ is a bounded function. Therefore, $K^*(g, g')$ has only a discrete spectrum in $\mathfrak{H}_0$, and since the spectrum of K^* in $\mathfrak{H}_0$ is the same as that of K because of the vanishing of the integral displayed above, we see that T_F has only a discrete spectrum in $\mathfrak{H}_0$.

To get a more detailed analysis of the structure of $\mathfrak{H}_0$, it is necessary to generalize the notion of point pair invariant introduced in §1.3. Consider the set $\mathscr{H}^n = \mathscr{H}^n(G, K)$ of all C^∞ functions of F of compact support on G satisfying $F(k(\theta)gk(\theta')) = e\left(\frac{n\theta}{2\pi}\right)F(g)e\left(\frac{n\theta'}{2\pi}\right)$, where n is an even integer. The set $\mathscr{H}^n(G, K)$ is a linear space over $\boldsymbol{C}$. If the product of the elements F_1 and F_2 of $\mathscr{H}^n$ is defined as the convolution $F_1 * F_2$, then the space $\mathscr{H}^n$ becomes an algebra over $\boldsymbol{C}$. This will be called the *Hecke algebra* of *weight* n with respect to G and K. An element $F \in \mathrm{M}^n$ defines a linear mapping T_F of the functions f of suitable integrability class on $G/\{\pm 1\}$ by $T_F f = f * F$. This mapping T_F will be called the *Hecke operator* determined by F, although it is the same thing as the operator which was defined above and denoted in the same way. If in particular the weight n is 0, then the Hecke operators are invariant operators in the sense of §1.3 defined by point pair invariants. In the same way as in the proof of Theorem 1.3.1, we can prove that all $\mathrm{M}^n(G, K)$ are commutative algebras, and as in the proof of Theorem 1.3.2 one may show that a cusp form in $\mathfrak{H}_{0,n}$ is an eigenfunction of every Hecke operator determined by some $F \in \mathrm{M}^n$. These special results lie within the scope of a general theory of Selberg on weakly symmetric Riemannian spaces (*cf.* p. 68). Thus, as in Theorem 5.2.2, it may be shown that $\mathfrak{H}_{0,n}$ is spanned by cusp forms.

The adjoint operator of the operator corresponding to $k(g, g') = F(g'^{-1}g)$ is given by $\overline{k(g', g)}$. Hence, the adjoint of the Hecke operator T_F is the Hecke operator $T_F{}^*$ with $F^*(g) = \overline{F(g^{-1})}$. Therefore, the adjoint of every Hecke operator coming from an element of M^n comes itself from an element of M^n, and consequently every Hecke operator is a normal operator. Summarizing these facts, we have

THEOREM 7.5.3. The space $\mathfrak{H}_0$ is the closure of the space spanned by the cusp forms, and a Hecke operator has a purely discrete spectrum in $\mathfrak{H}_0$.

The following theorem is essential for the spectral decomposition of $\mathfrak{H}_0$ in connection with the regular representation T_g of G.

THEOREM 7.5.4. If f is a cusp form in $\mathfrak{H}_{0,n}$, $(f \neq 0)$, then the space $\mathfrak{H}(f)$ generated by $T_g f$ for all $g \in G$ is an irreducible space with respect to the regular representation.

Proof. For any function $\xi \in \mathfrak{H}$, put

$$(P\xi)(g) = \frac{1}{2\pi}\int_0^{2\pi} \xi(gk(\theta))e\left(-\frac{n\theta}{2\pi}\right)d\theta$$

This operator satisfies $P^2 = P$, and is therefore a projection. Denote by $\mathfrak{H}_1$ the largest subspace of $\mathfrak{H}(f)$ consisting of functions ξ satisfying $P\xi = 0$ and invariant under all T_g. Then, the orthogonal complement $\mathfrak{H}_2$ of $\mathfrak{H}_1$ in $\mathfrak{H}(f)$ is also T_g = invariant. Since, however, $Pf = f$, the space $\mathfrak{H}_1$ does not coincide with $\mathfrak{H}(f)$, and consequently $\mathfrak{H}_2 \neq 0$. Take any $\xi \neq 0$ from $\mathfrak{H}_2$; then there exists a $g_0 \in G$ such that $PT_{g_0}\xi \neq 0$. Since ξ is a linear combination of functions of the form $T_{g'}f$, $(g' \in G)$, $T_{g_0}\xi$ is also a similar linear combination. But, it follows from the definition that

$$(PT_{g'}f)(g) = \frac{1}{2\pi}\int_0^{2\pi} f(gk(\theta)g')e\left(-\frac{n\theta}{2\pi}\right)d\theta$$

$$= \frac{1}{4\pi^2}\int_0^{2\pi}\int_0^{2\pi} f(gk(\theta)g'k(\theta'))e\left(-\frac{n\theta}{2\pi}\right)e\left(-\frac{n\theta'}{2\pi}\right)d\theta d\theta'.$$

This implies that $PT_{g'}f$ is the convolution product $f*F$, where F is the function which takes the value $e\left(-\frac{n\theta}{2\pi}\right)e\left(-\frac{n\theta'}{2\pi}\right)$ at $k(-\theta')g'^{-1}k(-\theta)$ and is equal to zero everywhere else. The function F does not belong to the Hecke algebra $\mathscr{H}^n(G, K)$ but is a limit of functions in $\mathscr{H}^n(G,K)$. Thus, the mapping $f \rightarrow f^*F$ is a limit of mappings given by Hecke operators. Hence we see that $\mathfrak{H}_2 \in PT_{g_0}\xi = \alpha f$ with $\alpha \in \mathbf{C}$, $\alpha \neq 0$. This implies that $\mathfrak{H}_2 = \mathfrak{H}(f)$, and at the same time that $\mathfrak{H}(f)$ is irreducible. *q.e.d.*

The projection P used in this theorem can be regarded as the unit element of $\mathscr{H}^n(G,K)$. As a matter of fact, P does not belong to $\mathscr{H}^n$ but can be interpreted as an element of a natural completion of $\mathscr{H}^n$.

Theorem 7.5.3 and Theorem 7.5.4 show that $\mathfrak{H}_0$ is an orthogonal sum of spaces of the form $\mathfrak{H}(f)$ defined in Theorem 7.5.4. Therefore, the regular representation restricted to $\mathbf{A}_0$ is decomposed into a discrete sum of irreducible representations. However, in the orthogonal decomposition

$$\mathfrak{H}_0 = \bigoplus_f \mathfrak{H}(f), \tag{7.5.2}$$

f does not range over a maximal set of linearly independent cusp forms; it is possible for $\mathfrak{H}(f) = \mathfrak{H}(f')$ to hold for two linearly independent cusp forms. A typical decomposition is obtained when f in (7.5.2) ranges over a set consisting of i) cusp forms in $\mathfrak{H}_{0,0}$, ii) those cusp forms in $\mathfrak{H}_{0,n}$, $(n < 0)$, which correspond in virtue of Theorem 7.5.1 to holomorphic automorphic forms on the upper half plane, and iii) their complex conjugates of the latter. However, a detailed investigation in this direction will not be given in this book. The irreducible representations of G given by $\mathfrak{H}(f)$ for which f is as described in ii) or iii) above belong to a particular class of representations of G called the *discrete series*.

At the end of this §, we propose to investigate a decomposition of the space Θ. We have already introduced in §7.4 the subspace $\hat{\Theta}_0$ of Θ defined as the closed subspace spanned by all incomplete theta series θ_ϕ for which the Eisenstein series $E_\phi(g, s)$ is holomorphic in Re $s \geqq \frac{1}{2}$. Now, we define another subspace $\hat{\Theta}$ of Θ to be the closed subspace spanned by the residues of $E_\phi(g, s)$ at its poles on $\frac{1}{2} < s \leqq 1$. Here, ϕ may be arbitrary except that it should, of course, satisfy some analytic conditions in order that $E_\phi(g, s)$ be well-defined; and the residue of $E_\phi(g, s)$ at $s_0 \in (\frac{1}{2}, 1]$ means the coefficient of $(s - s_0)^{-1}$ in the Laurent expansion of $E_\phi(g, s)$ in $s - s_0$, viewed as a function of g.

THEOREM 7.5.5. The space $\hat{\Theta}$ is the orthogonal complement of Θ_0 in Θ.

Proof. By Theorem 7.4.1, $E_\phi(g,s)$ is holomorphic in Re $s \geqq \frac{1}{2}$ if and only if $\hat{\theta}_\phi(g, s)$ has the same property. On the other hand, Theorem 7.2.2 gives

$$(\theta_\phi, E_{\phi'}(\bar{s})) = \int_K \hat{\theta}_\phi(k, s)\overline{L_{\phi'}(k, \bar{s})}d\theta \,.$$

So, $\lim(s - s_0)(\theta_\phi, E_{\phi'}(\bar{s})) = 0$, $(s \to s_0)$, whenever $\theta_\phi \in \Theta_0$. Conversely, this limit is non-zero for some ϕ' if $\hat{\theta}_\phi(k, s)$ has a pole at s_0. *q.e.d.*

For the arguments in this proof, it is enough to consider only those ϕ and ϕ' which are, for some n, in the space $\mathfrak{H}_{0,n}$, defined just before the statement of Theorem 7.5.1.

Theorem 7.5.5 tells us at the same time that the space Θ_0 in Theorem 5.2.4 can also be defined directly as the space of $\theta_\phi(z)$ such that $L_\phi(s)$ has a zero at each pole of $E(z, s)$ on $\frac{1}{2} \leqq s < 1$, notations being as in §3.1.

For an even integer n, denote by P_n the operator given by

$$(P_n f)(g) = \frac{1}{2\pi}\int_0^{2\pi} f(gk(\theta))e\left(-\frac{n\theta}{2\pi}\right)d\theta$$

for $f \in \mathfrak{H}$, i.e., the same operator as P in Theorem 7.5.4. The image $P_n f$ then has the property $(P_n f)(gk(\theta)) = (P_n f)(g) \cdot e\left(\frac{n\theta}{2\pi}\right)$, and $P_n^2 = P_n$.

THEOREM 7.5.6. Let Γ be a discontinuous group of finite type reduced at ∞ which is assumed to be the only cusp up to equivalence, and let $E(g, s)_n$ be the Eisenstein series defined in §6.1. Then, for any $g_0 \in G = SL(2, \boldsymbol{R})$, there exists a constant c_{g_0} depending on g_0 and on s such that

$$P_n T_{g_0} E(g, s)_0 = c_{g_0} E(g, s)_n \,.$$

As a function of $g_0 \in G$, the constant c_{g_0} is not identically 0 for any fixed s, (Re $s > 0$).

Proof. The equality in the theorem is equivalent to

$$P_n T_{g_0} y(g)^s = c_{g_0} y(g)^s e\left(\frac{n\theta}{2\pi}\right),$$

where $s \in \boldsymbol{C}$ and $y(g) = y$ if $g = n(x)a(y)k(\theta)$ is the Iwasawa decomposition of $g \in G$.

Now, by definition,

$$P_n T_{g_0} y(g)^s = \frac{1}{2\pi}\int_0^{2\pi} y(gk(\theta')g_0)^s e\left(-\frac{n\theta'}{2\pi}\right)d\theta'$$

If $g = n(x)a(y)k(\theta)$, then this integral is equal to

$$\frac{1}{2\pi} y(g)^s \int_0^{2\pi} y(k(\theta+\theta')g_0)^s e\left(-\frac{n\theta'}{2\pi}\right)d\theta'$$

$$= \frac{1}{2\pi} y(g)^s e\left(\frac{n\theta}{2\pi}\right)\int_0^{2\pi} y(k(\theta')g_0)^s e\left(-\frac{n\theta'}{2\pi}\right)d\theta'.$$

Put

$$c_{g_0} = \frac{1}{2\pi}\int_0^{2\pi} y(k(\theta')g_0)^s e\left(-\frac{n\theta'}{2\pi}\right)d\theta';$$

then c_{g_0} depends only on g_0, and c_{g_0} is not identically zero for any fixed s, because, for $g_0 = \begin{pmatrix}\sqrt{y_0} & \\ & \sqrt{y_0}^{-1}\end{pmatrix}$, $(y_0 > 0)$, say, we have

$$2\pi c_{g_0} = \int_0^{2\pi} \frac{y_0^s}{|\sqrt{-1}y_0 \sin\theta' + \cos\theta'|^{2s}} e\left(-\frac{n\theta'}{2\pi}\right)d\theta'$$

$$= y_0^s \int_0^{2\pi} \frac{e\left(-\frac{n\theta'}{2\pi}\right)}{(1+(y_0^2-1)\sin^2\theta')^s} d\theta'$$

$$= (2y_0)^s \int_0^{2\pi} \frac{\cos n\theta'}{(1+y_0^2+(1-y_0^2)\cos 2\theta')^s} d\theta',$$

and this can be expressed in terms of a hypergeometric function, which is not identically zero. *q.e.d.*

The space $\hat{\Theta} \in \mathfrak{H} = L^2(\Gamma\backslash G)$ is in general not finite dimensional over $\boldsymbol{C}$, because the residues of $E(g, s)_n$, $(n = 0, \pm 2, \pm 4, \ldots)$, at a pole $s_0 \in (\frac{1}{2}, 1]$ of $E(z, s) = E(g, s)_0$ can be linearly independent elements of $\hat{\Theta}$. If $s_0 \neq 1$, then they are actually linearly independent as one can see from Theorem

6.1.1. On the other hand, Theorem 7.5.6 shows that for any n, $E(g, s)_n$ is contained in the space generated by all $T_{g_0}E(g, s)_0$, $(g_0 \in G)$.

Thus we have

THEOREM 7.5.7. The space $\hat{\Theta}$ is the orthogonal sum of a finite number of irreducible spaces of regular representations.

The poles of $E(z, s)$ in $(\frac{1}{2}, 1]$ are poles of the function $\varphi(s)$ appearing in the constant term of the Fourier expansion of $E(z, s)$. Each irreducible component in Theorem 7.5.7 corresponds to a pole s_0 of $\varphi(s)$ with $\frac{1}{2} < s_0 \leqq 1$. If $s_0 = 1$, then the corresponding space consists merely of scalar functions. So, if $\varphi(s)$ has no pole other than $s_0 = 1$, then $\hat{\Theta} \cong C$ and $\hat{\Theta}$ is one dimensional. In all other cases $\dim_C \hat{\Theta} = \infty$.

The spectral decomposition of the space $\mathfrak{H} = L^2(\Gamma\backslash G)$ for a general discontinuous group Γ of finite type is not very much different from that given by the above results.

In the present instance, the space Θ is the closed subspace of $\mathfrak{H}$ spanned by all $\theta_{i,\varphi}$ as defined in (7.4.1). The space Θ_0 is the closed space spanned by all $\theta_{i,\varphi}$ for which $\hat{\theta}_{ij,\varphi}(g, s)$ in (7.4.2) is holomorphic in $\frac{1}{2} \leqq \operatorname{Re} s$ for every $j = 1, 2, \ldots, h$. The space $\mathfrak{H}_0$ is the space of all $f \in \mathfrak{H}$ such that the constant term of the Fourier expansion of f is zero at every cusp of Γ, i.e., such that $\int_0^1 f(\sigma_i n(x)g)dx = 0$ for $i = 1, 2, \ldots, h$, the notations being as in (7.4.1). Furthermore, a cusp form means a function $f \in \mathfrak{H}_0$ which is an eigenfunction of the Casimir operator and of $\frac{\partial}{\partial\theta}$, and the space $\hat{\Theta}$ is the closed space spanned by all the residues at their poles on $\frac{1}{2} < s \leqq 1$ of the function $E_{i,\varphi}(g, s)$ in (6.2.9), $(i = 1, 2, \ldots, h)$.

The space $\mathfrak{H}_0$ is the orthogonal complement of Θ in $\mathfrak{H}$, and the space $\hat{\Theta}$ is the orthogonal complement of Θ_0 in Θ; these are analogues of Theorem 7.5.2 and Theorem 7.5.5. A kernel function which generalizes the function $H(g, g')$ in (7.5.1) can be defined in fashion similar to that for $H(z, z')$ in (5.3.6), and it may be proved that $\mathfrak{H}_0$ is spanned by cusp forms. This is an analogue of Theorem 7.5.3.

The decomposition of the regular representation $T_{g_0} : f(g) \to f(gg_0)$ is also similar to that in the special case treated before. Because of the Plancherel formula (7.4.3), T_{g_0} is on Θ_0 an h-dimensional continuous sum of irreducible representations in principal series, T_{g_0} is on $\mathfrak{H}_0$ a discrete sum of infinitely many irreducible representations determined by cusp forms according to Theorem 7.5.4, and on $\hat{\Theta}$, T_{g_0} is a finite sum of irreducible representations as may be seen from an analogue of Theorem 7.5.7.

Appendix

Explicit trace formulas

In this appendix we denote by Γ a discontinuous group of finite type acting on the upper half plane H, and, except in some comments at the end, we assume that Γ is reduced at ∞ which is assumed to be the only cusp of Γ up to equivalence. In Chap. VI and Chap. VII, we have taken a discontinuous group to be a subgroup of $G = SL(2, \boldsymbol{R})$, but in this appendix, a discontinuous group will again stand for a group of fractional linear transformations, i.e., a subgroup of $G/\{\pm 1\}$.

Let $k(z, z')$ be a point pair invariant, and let $\hat{K}(z, z') = \sum_{\sigma \in \Gamma} k(z, \sigma z') - \hat{H}(z, z')$ be the kernel function, defined as in Theorem 5.3.3, which determines an integral operator $f \to \int_{\mathscr{D}} \hat{K}(z,z')f(z')dz'$ on the space of functions on $\mathscr{D} = \Gamma \backslash H$. The constant $\int_{\mathscr{D}} \hat{K}(z, z)dz$ is then called the *trace* of the integral operator. It has been shown by Selberg that this trace may be reduced to the sum of components corresponding to conjugacy classes of Γ, and that each component of the trace can be expressed by a simple and smooth formula containing the Selberg transform h of k as defined in §5.3. This appendix is devoted to the proof of these results.

Denote by $\{\tau\}$ the conjugacy class of $\tau \in \Gamma$. Then, formally at first, we have

$$\int_{\mathscr{D}} K(z, z)dz = \sum_{\{\tau\}} \sum_{\sigma \in \{\tau\}} \int_{\mathscr{D}} k(z, \sigma z)dz.$$

Denote now by Γ_τ the centralizer of τ in Γ, and put $\mathscr{D}_\tau = \Gamma_\tau \backslash H$. Then, it is easy to see that

$$\sum_{\sigma \in \{\tau\}} \int_{\mathscr{D}} k(z, \sigma z)dz = \int_{\mathscr{D}_\tau} k(z, \tau z)dz.$$

If τ is not a parabolic transformation, then the *component*

$$c(\tau) = \int_{\mathscr{D}_\tau} k(z, \tau z)dz$$

of the trace is expressible in terms of $h(s)$. For a parabolic element τ the individual $c(\tau)$ do not exist in general, so we consider a modified *component*

$$\text{(A. 1)} \quad c(\infty) = \lim_{Y\to\infty}\left[\int_0^Y\int_0^1 \sum_{1\neq\gamma_0\in\Gamma_0} k(z, \gamma_0 z)dz - \int_{\mathscr{D}_Y} \hat{H}(z, z)dz\right]$$

for the cusp ∞, where $\mathscr{D}_Y$ is as in Theorem 2.3.3. Roughly speaking, this corresponds to the difference

$$\sum_{\gamma\in\Gamma_0}\int_{\mathscr{D}_\gamma} k(z, \gamma z)dz - \int_{\mathscr{D}} \hat{H}(z, z)dz$$

with $\Gamma_0 = \{n(b)|b \in \boldsymbol{Z}\}$, $\left(n(x) = \begin{pmatrix}1 & x\\ & 1\end{pmatrix}\right)$. We have, finally, the formula

$$\text{(A. 2)} \quad \int_{\mathscr{D}} \hat{K}(z, z)dz = c(I) + \sum_{\{R\}} c(R) + \sum_{\{P\}} c(P) + c(\infty),$$

where I, R, and P denote respectively the identity, an elliptic transformation, and an hyperbolic transformation in Γ.

The trace in this formula for the integral operator also has an interpretation as a sum of eigenvalues. To explain this situation briefly, recall the orthogonal decomposition of $\mathfrak{H} = L^2(\Gamma\backslash H)$ into the sum $\Theta_0 \oplus \hat{\Theta} \oplus \mathfrak{H}_0$ as given in Theorem 5.2.4, where $\Theta_0 \oplus \hat{\Theta} = \Theta$ is the space of incomplete theta series and $\mathfrak{H}_0$ is the space spanned by cusp forms. Then, for any $\theta_\phi(z) = \sum_{\sigma\in\Gamma_0\backslash\Gamma} \phi(\sigma z) \in \Theta$, we have

$$\int_{\mathscr{D}} K(z, z')\theta_\phi(z')dz' = \int_H k(z, z')\theta_\phi(z')dz'$$

$$= \int_H k(z, z') \sum_{\sigma\in\Gamma_0\backslash\Gamma} \phi(\sigma z')dz' = \sum_{\sigma\in\Gamma_0\backslash\Gamma}\int_H k(\sigma z, z')\phi(z')dz' = \theta_{\phi'}(z) \in \Theta$$

with

$$\text{(A. 3)} \quad \phi'(z) = \int_H k(z, z')\phi(z')dz'.$$

On the other hand,

$$\int_{\mathscr{D}} \hat{H}(z, z')\theta_\phi(z')dz' = \int_{\mathscr{D}} \frac{1}{4\pi}\int_{-\infty}^{\infty} h(r)E(z, s)E(z', \bar{s})dr\theta_\phi(z')dz'$$

for $s = \frac{1}{2} + \sqrt{-1}r$, and

$$\int_{\mathscr{D}} E(z', \bar{s})\theta_\phi(z')dz' = \int_{\Gamma_0\backslash H} E(z', \bar{s})\phi(z')dz'$$
$$= \int_0^\infty \psi(y')y'^{s-1}\frac{dy'}{y'} + \varphi(\bar{s})\int_0^\infty \psi(y')y'^{-s}\frac{dy'}{y'}$$
$$= L_\phi(s) + \varphi(1-s)L_\phi(1-s),$$

where $\psi(y') = \phi(a(y')\sqrt{-1})$ with $a(y') = \begin{pmatrix} \sqrt{y'} & \\ & \sqrt{y'^{-1}} \end{pmatrix}$, and L_ϕ is as in §3.1 or in (6.2.4). Thus, from the functional equation of $E(z, s)$ and from (6.2.5), one has

$$\int_{\mathscr{D}} \hat{H}(z, z')\theta_\phi(z')dz' = \frac{1}{4\pi}\int_{-\infty}^\infty h(r)E(z, s)(L_\phi(s) + \varphi(1-s)L_\phi(1-s))dr$$
$$= \frac{1}{2\pi}\int_{-\infty}^\infty h(r)L_\phi(s)E(z, s)dr = \frac{1}{2\pi\sqrt{-1}}\int_{\operatorname{Re} s=1/2} h(s)E_\phi(z, s)ds.$$

Assume now that $\theta_\phi \in \Theta_0$, or equivalently, because of Theorem 7.4.1, that $E_\phi(z, s)$ has no pole on $\frac{1}{2} \leqq s \leqq 1$; then the above formula is equal to

$$\frac{1}{2\pi\sqrt{-1}}\int_{\operatorname{Re} s=S>1} h(s)E_\phi(z, s)ds.$$

On the other hand, if ϕ' is as in (A. 3) and $y' = y(z') = \operatorname{Im} z'$, then

$$L_{\phi'}(z, s) = \int_0^\infty \int_H k(a(y)z, z')\phi(z')dz'y^{-s}\frac{dy}{y}$$
$$= \int_0^\infty \int_H k(z, z')\phi(yy')dz'y^{-s}\frac{dy}{y} = \int_H k(z, z')y'^s dz' \cdot \int_0^\infty \psi(y)y^{-s}\frac{dy}{y}$$
$$= h(s)L_\phi(s)y^s,$$

so that (7.4.4) implies

$$\frac{1}{2\pi\sqrt{-1}}\int_{\operatorname{Re} s=S>1} h(s)E_\phi(z, s)ds = \theta_{\phi'}(z).$$

Thus, we have

$$\int_{\mathscr{D}} \hat{H}(z, z')\theta_\phi(z')dz' = \theta_{\phi'}(z),$$

and

$$\int_{\mathscr{D}} \hat{H}(z, z')\theta_\phi(z')dz' = 0.$$

Therefore, the integral operator defined by the kernel function $\hat{K}$ becomes the zero operator on Θ_0. On $\hat{\Theta} \oplus \mathfrak{H}_0$, the integral operator defined by $\hat{K}$ has the same effect as the integral operator defined by K. To see this, it is enough to show that $(E(z, s), f) = 0$ for any $f \in \hat{\Theta} \oplus \mathfrak{H}_0$ and for $\operatorname{Re} s = 1/2$. This orthogonality can be verified either by Theorem 2.3.1 or simply from the fact that the Laplacian D is self-adjoint. Namely, the equation $s(s-1)(E(z, s), f) = (DE(z, s), f) = (E(z, s), Df) = \bar{\lambda}_0(E(z, s), f)$ follows from $Df = \lambda_0 f$; since $s(s-1)$ is not a constant, $(E(z, s), f) = 0$ must hold.

Let now $F_1, F_2, \ldots$ be an orthonormal system of automorphic functions generating $\hat{\Theta} \oplus \mathfrak{H}_0$ and satisfying $DF_i = \lambda_i F_i$. Then,

$$\int_{\mathscr{D}} \hat{K}(z, z')F_i(z')dz' = h(r_i)F_i(z)$$

with $\lambda_i = -\dfrac{1}{4} - r_i^2$ by the arguments in §5.3. So, under certain conditions, for example whenever $K(z, z')$ is of Hilbert-Schmidt type on $\mathfrak{H}_0$, we have

$$\hat{K}(z, z') = \sum h(r_i)F_i(z)\overline{F_i(z')},$$

and consequently

$$\int_{\mathscr{D}} \hat{K}(z, z)dz = \frac{1}{2}\sum h(r_i),$$

where $\lambda_i = s_i(s_i - 1)$ with $s_i = 1/2 + \sqrt{-1}r_i$, and $\operatorname{Re} s_i = 1/2$ or $0 \leqq s_i \leqq 1$. This explains the meaning of the trace as the sum of eigenvalues.

We now propose to study the components $c(I)$, $c(R)$, and $c(P)$ of traces appearing in (A. 2), and to calculate final and satisfactory expressions for them.

As for $c(I)$, it follows from the definition that $\Gamma_I = \Gamma$. Therefore, we have

$$c(I) = \int_{\mathscr{D}} k(0)dz = k(0)A(\mathscr{D}),$$

where $A(\mathscr{D})$ is the measure of the fundamental domain $\mathscr{D}$.

As for $c(R)$, we note first Γ_R is a cyclic subgroup of finite order in Γ generated by an elliptic element $R_0 \in \Gamma$. This is easy to see if one considers the Jordan canonical form. The element R_0 may be so chosen that the angle $\theta_0 = \pi/m$, for which R_0 is conjugate in $G = SL(2, R)$ to $\begin{pmatrix} \cos\theta_0 & -\sin\theta_0 \\ \sin\theta_0 & \cos\theta_0 \end{pmatrix}$, is the smallest positive angle among all such angles determined by the elements of the group generated by R_0. Such an R_0 is called a *primitive* elliptic element of Γ. Every elliptic element in Γ is a power of a uniquely determined primitive one. Assume that $R = R_0^l$ with primitive R_0, and let m be the order of R_0. Then, putting $\begin{pmatrix} \cos l\theta_0 & -\sin l\theta_0 \\ \sin l\theta_0 & \cos l\theta_0 \end{pmatrix} = \begin{pmatrix} \alpha & -\beta \\ \beta & \alpha \end{pmatrix}$, we have

$$c(R) = \int_{\mathscr{D}_{R_0}} k\left(z, \frac{\alpha z - \beta}{\beta z + \alpha}\right) dz = \frac{1}{m} \int_H k\left(\frac{\left|z - \dfrac{\alpha z - \beta}{\beta z + \alpha}\right|^2}{y \cdot \dfrac{y}{|\beta z + \alpha|^2}}\right) \frac{dxdy}{y^2}$$

$$= \frac{1}{m} \int_0^\infty \int_{-\infty}^\infty k\left(\frac{|\beta z^2 + \beta|^2}{y^2}\right) \frac{dxdy}{y^2} = \frac{1}{m} \int_0^\infty \int_{-\infty}^\infty k\left(\frac{|z^2 + 1|^2}{y^2} \sin^2 \frac{l\pi}{m}\right) \frac{dxdy}{y^2}.$$

As for $c(P)$, we note that the Jordan canonical form of P is $\begin{pmatrix} \rho & \\ & \rho^{-1} \end{pmatrix}$ with $\rho > 1$. The number ρ^2 will be called the *norm* of P and will be denoted by NP. Using the Jordan canonical form we can also conclude that Γ_P is an infinite cyclic group generated by a hyperbolic element $P_0 \in \Gamma$ which is not a power of any element in Γ_P except $P_0^{\pm 1}$. Such a P_0 will be called a *primitive* hyperbolic element of Γ. Every hyperbolic element in Γ is a power of a primitive hyperbolic element which is uniquely determined up to inversion. Assume $P = P_0^l$ with primitive P_0, and put $\rho^2 = NP_0$. Then,

$$c(P) = \int_{\mathscr{D}_{P_0}} k(z, \rho^{2l} z) dz$$

$$= \int_1^{\rho^2} \int_{-\infty}^\infty k\left(\frac{|z - \rho^{2l} z|^2}{\rho^{2l} y^2}\right) \frac{dxdy}{y^2} = \int_1^{NP_0} \int_{-\infty}^\infty k\left(\frac{(NP_0 - 1)^2}{NP_0^l} \frac{|z|^2}{y^2}\right) dxdy.$$

What we have to do next is to express the above integral formulas for $c(I)$, $c(R)$, and $c(P)$ in terms of the Selberg transform h of k.

By Theorem 5.3.1,

$$k(0) = -\frac{1}{\pi} \int_0^\infty \frac{dQ(w)}{\sqrt{w}} = -\frac{1}{\pi} \int_0^\infty \frac{g'(u)}{e^{u/2} - e^{-u/2}} du$$

$$= \frac{1}{2\pi^2} \int_{-\infty}^\infty \int_0^\infty r h(r) \frac{\sin ru}{e^{u/2} - e^{-u/2}} dudr.$$

Form the formula

$$\int_0^\infty \operatorname{cosech} \frac{u}{2} \sin ru \, du = \pi \tanh \pi r$$

for the Fourier transformation, we have

$$k(0) = \frac{1}{4\pi}\int_{-\infty}^{\infty} rh(r)\frac{e^{\pi r} - e^{-\pi r}}{e^{\pi r} + e^{-\pi r}}dr.$$

Thus,

$$c(I) = \frac{A(\mathscr{D})}{4\pi}\int_{-\infty}^{\infty} r\frac{e^{\pi r} - e^{-\pi r}}{e^{\pi r} + e^{-\pi r}}h(r)dr. \tag{A.4}$$

The computation of $c(R)$ requires more complicated arguments. Put first $\beta = \sin\frac{l\pi}{m}$; then

$$c(R) = \frac{1}{m}\int_0^{\infty}\int_{-\infty}^{\infty} k\left(\beta^2\left(\frac{(x^2+1)^2}{y^2} + 2(x^2-1) + y^2\right)\right)\frac{dxdy}{y^2}.$$

Change the variable x here to $t = \beta^2\left(\frac{(x^2+1)^2}{y^2} + 2(x^2-1) + y^2\right)$; then $dt = 4\beta^2\left(\frac{x^2+1}{y^2} + 1\right)xdx = \frac{4\beta^2}{y}\sqrt{4 + \frac{t}{\beta^2}}\,xdx$, $\min\limits_{x\in R} t = t\Big|_{x=0} = \beta^2\left(y - \frac{1}{y}\right)^2$, and $x^2 = (\alpha_2 - y)(y - \alpha_1)$, where $\alpha_1, \alpha_2, (\alpha_1 < \alpha_2)$, are the positive roots of $t = \beta^2\left(y - \frac{1}{y}\right)^2$. Then the above integral for $c(R)$ is equal to

$$\begin{aligned}
&\frac{2}{m}\int_0^{\infty}\int_{\alpha_1}^{\alpha_2} k(t)\frac{\partial x}{\partial t}\frac{dydt}{y^2}\\
&= \frac{1}{2\beta^2 m}\int_0^{\infty}\frac{k(t)}{\sqrt{4 + \frac{t}{\beta^2}}}dt\int_{\alpha_1}^{\alpha_2}\frac{dy}{y\sqrt{(\alpha_2 - y)(y - \alpha_1)}}\\
&= \frac{\pi}{2m\beta}\int_0^{\infty}\frac{k(t)}{\sqrt{4\beta^2 + t}}dt
\end{aligned}$$

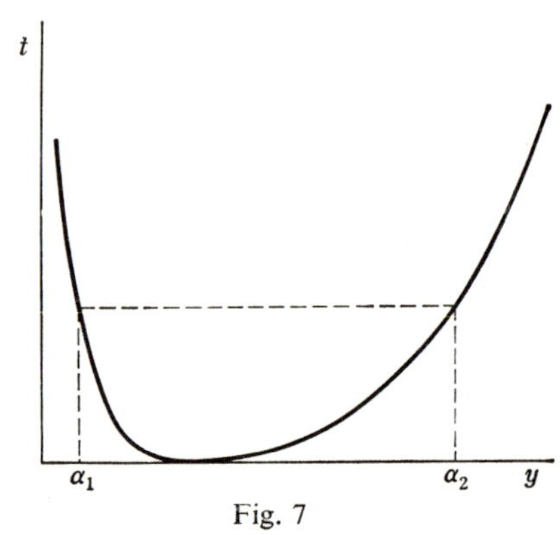

Fig. 7

because of $\alpha_1\alpha_2 = 1$, (*cf.* Fig. 7). Now using the formula in Theorem 5.3.1, and applying partial differentiation we get

$$\frac{\pi}{2m\beta}\int_0^\infty \frac{k(t)}{\sqrt{4\beta^2+t}}dt$$

$$= \frac{\pi}{2m\beta}\int_0^\infty \frac{1}{\sqrt{4\beta^2+t}}\frac{-1}{\pi}\int_t^\infty \frac{dQ(w)}{\sqrt{w-t}}dt$$

$$= -\frac{1}{2m\beta}\int_0^\infty\int_t^\infty \frac{1}{\sqrt{4\beta^2+t}}\frac{dQ(w)}{\sqrt{w-t}}dt$$

$$= -\frac{1}{2m\beta}\int_0^\infty\int_0^w \frac{dt}{\sqrt{(4\beta^2+t)(w-t)}}dQ(w)$$

$$= \frac{1}{2m\beta}\int_0^\infty Q(w)d\left[\int_0^w \frac{dt}{\sqrt{(4\beta^2+t)(w-t)}}\right]$$

$$= \frac{1}{2m\beta}\int_0^\infty Q(w)d\left[\frac{\pi}{2}+\arcsin\frac{w-4\beta^2}{w+4\beta^2}\right] = \frac{1}{2m\beta}\int_0^\infty Q(w)\frac{2\beta}{\sqrt{w}(w+4\beta^2)}dw.$$

This may be transformed further as follows:

$$\frac{1}{m}\int_0^\infty g(u)\frac{e^{u/2}+e^{-u/2}}{(e^{u/2}-e^{-u/2})^2+4\beta^2}du$$

$$= \frac{1}{2\pi m}\int_0^\infty \frac{e^{u/2}+e^{-u/2}}{(e^{u/2}-e^{-u/2})^2+4\beta^2}\int_{-\infty}^\infty h(r)e\left(\frac{ru}{2\pi}\right)drdu$$

$$= \frac{1}{2\pi m}\int_{-\infty}^\infty h(r)\int_0^\infty \frac{e^{u/2}+e^{-u/2}}{(e^{u/2}+e^{-u/2})^2+4\beta^2}e\left(\frac{ru}{2\pi}\right)dudr$$

$$= \frac{1}{4\pi m}\int_{-\infty}^\infty h(r)\int_{-\infty}^\infty \frac{e^{u/2}+e^{-u/2}}{(e^{u/2}-e^{-u/2})^2+4\beta^2}e\left(\frac{ru}{2\pi}\right)dudr,$$

since $h(r)$ is an even function. The Fourier integral in the last formula is computed by calculating the residues of the integrand. Since the solutions u_0 of the equation $(e^{u/2}-e^{-u/2})^2 = -4\beta^2 = (e^{l\pi\sqrt{-1}/m}-e^{-l\pi\sqrt{-1}/m})^2$ are given by $u_0 = \left(\pm\frac{2l\pi}{m}+2n\pi\right)\sqrt{-1}$, the poles of the function

$$\frac{e^{u/2}+e^{-u/2}}{(e^{u/2}-e^{-u/2})^2+4\beta^2}e\left(\frac{ru}{2\pi}\right)$$

on the upper half plane $\operatorname{Im} u>0$ are $\left(\frac{2l\pi}{m}+2n\pi\right)\sqrt{-1}$, $(n = 0, 1, 2, \ldots)$, or $\left(-\frac{2l\pi}{m}+2n\pi\right)\sqrt{-1}$, $(n = 1, 2, \ldots)$, and the residue of the above function of u at u_0 is

$$\frac{e^{-(\pm 2l\pi/m+2n\pi)r}}{2\sqrt{-1}\cdot(-1)^n\sin\dfrac{\pm l\pi}{m}}=\pm\frac{e^{(\mp 2l\pi/m-2n\pi)r}}{2\sqrt{-1}(-1)^n\beta}$$

Using these results, we have

$$\int_{-\infty}^{\infty}\frac{e^{u/2}+e^{-u/2}}{(e^{u/2}-e^{-u/2})^2+4\beta^2}e\left(\frac{ru}{2\pi}\right)du=\frac{\pi}{\beta}e^{-(2l\pi/m)r}(1-e^{-2\pi r}+e^{-4\pi r}-\dots)$$
$$+\frac{\pi}{\beta}e^{(2l\pi/m)r}(-e^{-2\pi r}+e^{-4\pi r}-\dots)=\frac{\pi}{\beta}\frac{e^{-(2l\pi/m)r}+e^{-(2\pi-2l\pi/m)r}}{1+e^{-2\pi r}}.$$

Since $h(r)$ is an even function, this implies that

$$\frac{1}{4\pi m}\int_{-\infty}^{\infty}h(r)\int_{-\infty}^{\infty}\frac{e^{u/2}+e^{-u/2}}{(e^{u/2}-e^{-u/2})^2+4\beta^2}e\left(\frac{ru}{2\pi}\right)dudr$$
$$=\frac{1}{2m\beta}\int_{-\infty}^{\infty}\frac{e^{-(2l\pi/m)r}}{1+e^{-2\pi r}}h(r)dr.$$

Thus the final formula is

(A. 5) $$c(R)=\frac{1}{2m\sin\dfrac{l\pi}{m}}\int_{-\infty}^{\infty}\frac{e^{-(2l\pi/m)r}}{1+e^{-2\pi r}}h(r)dr.$$

The corresponding formula for $c(P)$ is given by

(A. 6) $$c(P)=\frac{\log NP_0}{NP_0{}^{l/2}-NP_0{}^{-l/2}}g(l\log NP_0).$$

The proof of this formula will be omitted, because it is simple.†

Now we come to the most complicated part of the explicit computation in regard to the trace $\int_{\mathscr{D}}\hat{K}(z,z)dz$. Namely, we now propose to treat the component $c(\infty)$ in (A. 1), and to find a satisfactory expression for it.

We observe first that

$$\sum_{1\neq\gamma_0\in\Gamma_0}\int_0^Y\int_0^1 k(z,\gamma_0 z)dz=\sum_{0\neq b\in Z}\int_0^Y k\left(\frac{b^2}{y^2}\right)\frac{dy}{y^2},$$

using notation as indicated in Theorem 5.3.1. Put

$$f(x)=2\int_0^Y k\left(\frac{x^2}{y^2}\right)\frac{dy}{y^2}=\frac{1}{x}\int_{x^2/Y^2}^{\infty}\frac{k(t)}{\sqrt{t}}dt.$$

† Proofs of the formulas (A. 4), (A. 5), and (A. 6) have been published by **M. Kuga**, *Sugaku* (Japanese), **9**–3, 166–185 (1958).

Then, what we want to get is $\sum_{b=1}^{\infty} f(b)$, and a well-known elementary summation formula yields

(A. 7)
$$\sum_{b=1}^{\infty} f(b) = \frac{1}{2} f(1) + \int_1^{\infty} f(x)dx + \int_1^{\infty} f'(x)\{x\}\,dx,$$

where $\{x\} = x - [x] - \frac{1}{2}$, and $[x]$ is the largest integer not exceeding $x \in \boldsymbol{R}$.

As for the first term on the right hand side of (A. 7), it is easy to see that

$$\frac{1}{2} f(1) = \frac{1}{2} \int_{1/Y^2}^{\infty} \frac{k(t)}{\sqrt{t}}\,dt \to \frac{1}{2} g(0) \qquad \text{as } Y \to \infty.$$

The second term requires more work. As a preliminary step, a straightforward calculation shows that

$$\int_1^{\infty} f(x)dx = \int_1^{\infty} \frac{1}{x} \int_{x^2/Y^2}^{\infty} \frac{k(t)}{\sqrt{t}}\,dtdx = \int_{1/Y}^{\infty} \int_{Y^2}^{\infty} \frac{k(t)}{\sqrt{t}}\,dt \frac{dy}{y}$$

$$= \int_0^Y \int_{1/y^2}^{\infty} \frac{k(t)}{\sqrt{t}} dt \frac{dy}{y} = \int_1^Y \int_{1/y^2}^{\infty} \frac{k(t)}{\sqrt{t}} dt \frac{dy}{y} + \int_0^1 \int_{1/y^2}^{\infty} \frac{k(t)}{\sqrt{t}} dt \frac{dy}{y}$$

$$= \int_1^Y g(0) \frac{dy}{y} - \int_1^Y \int_0^{1/y^2} \frac{k(t)}{\sqrt{t}}\,dt \frac{dy}{y} + \int_0^1 \int_{1/y^2}^{\infty} \frac{k(t)}{\sqrt{t}}\,dt \frac{dy}{y}$$

$$= g(0)\log Y + \int_0^1 \int_{1/y^2}^{\infty} \frac{k(t)}{\sqrt{t}} dt \frac{dy}{y} - \int_1^{\infty} \int_0^{1/y^2} \frac{k(t)}{\sqrt{t}} dt \frac{dy}{y} + \int_Y^{\infty} \int_0^{1/y^2} \frac{k(t)}{\sqrt{t}} dt \frac{dy}{y}.$$

Now, we have

$$\int_0^1 \int_{1/y^2}^{\infty} \frac{k(t)}{\sqrt{t}}\,dt \frac{dy}{y} - \int_1^{\infty} \int_0^{1/y^2} \frac{k(t)}{\sqrt{t}}\,dt \frac{dy}{y}$$

$$= \lim_{\varepsilon \to 0} \left\{ \int_{\varepsilon}^1 \int_0^{\infty} \frac{k(t)}{\sqrt{t}}\,dt \frac{dy}{y} - \int_{\varepsilon}^{\infty} \int_0^{1/y^2} \frac{k(t)}{\sqrt{t}}\,dt \frac{dy}{y} \right\}$$

$$= \lim_{\varepsilon \to 0} \left\{ - g(0)\log \varepsilon - \left(\log y \int_0^{1/y^2} \frac{k(t)}{\sqrt{t}}\,dt \right|_{\varepsilon}^{\infty} - \int_{\varepsilon}^{\infty} \log y \cdot \frac{k\left(\frac{1}{y^2}\right)}{\sqrt{\frac{1}{y^2}}} \left(-\frac{2}{y^3}\right) dy \right) \right\}$$

$$= \lim \left\{ - g(0)\log \varepsilon + \log \varepsilon \cdot \int_0^{1/\varepsilon^2} \frac{k(t)}{\sqrt{t}}\,dt - \int_{\varepsilon}^{\infty} \frac{2\log y\, k\left(\frac{1}{y^2}\right)}{y^2} dy \right\}$$

$$= \frac{1}{2}\int_0^\infty \frac{\log t \cdot k(t)}{\sqrt{t}}dt,$$

(*cf.* Fig. 8).

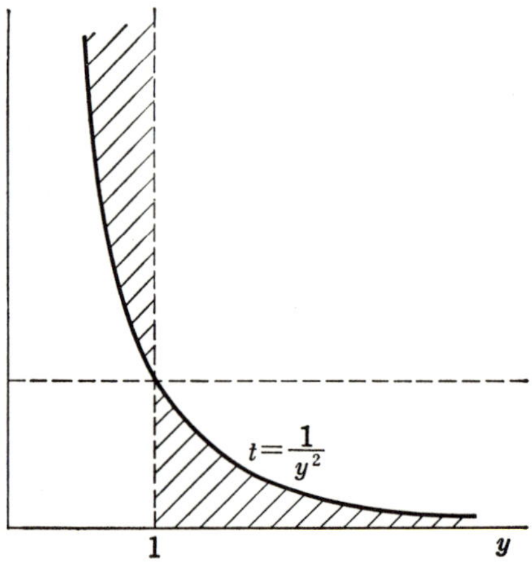

Fig. 8

Furthermore,

$$\int_0^\infty \frac{\log t \cdot k(t)}{\sqrt{t}}dt = -\frac{1}{\pi}\int_0^\infty\int_t^\infty \frac{\log t}{\sqrt{t}\,\sqrt{w-t}}Q'(w)dwdt$$

$$= -\frac{1}{\pi}\int_0^\infty\int_0^w \frac{\log t}{\sqrt{t}\,\sqrt{w-t}}Q'(w)dtdw$$

$$= -\frac{1}{\pi}\int_0^\infty\int_0^1 \frac{\log wt}{\sqrt{t}\,\sqrt{1-t}}dtQ'(w)dw = -\frac{1}{\pi}\int_0^\infty(\pi\log w + c_1)Q'(w)dw$$

with

$$c_1 = \int_0^1 \frac{\log t}{\sqrt{t}\,\sqrt{1-t}}dt = -2\pi\log 2\,,$$

and one has

$$-\frac{1}{\pi}\int_0^\infty(\pi\log w + c_1)Q'(w)dw = \lim_{\varepsilon\to 0}\left\{\left(-\log w - \frac{c_1}{\pi}\right)Q(w)\Big|_\varepsilon^\infty - \int_\varepsilon^\infty \frac{-Q(w)}{w}dw\right\}$$

$$= \frac{c_1}{\pi}g(0) + \lim_{\varepsilon\to 0}\left((\log\varepsilon \cdot Q(\varepsilon) + \int_\varepsilon^\infty \frac{Q(w)}{w}dw\right).$$

Putting here $\varepsilon = e^{\varepsilon'} + e^{-\varepsilon'} - 2$ with $\varepsilon' > 0$, one obtains

$$\int_{\varepsilon}^{\infty} \frac{Q(w)}{w} dw = \int_{\varepsilon'}^{\infty} \frac{g(u)}{e^u + e^{-u} - 2} (e^u - e^{-u}) du$$

$$= \int_{\varepsilon'}^{\infty} \frac{e^{u/2} + e^{-u/2}}{e^{u/2} - e^{-u/2}} \frac{1}{2\pi} \int_{-\infty}^{\infty} h(r) e\left(-\frac{ru}{2\pi}\right) dr du$$

$$= \frac{1}{2\pi} \int_{\varepsilon'}^{\infty} \int_{-\infty}^{\infty} \frac{1 + e^{-u}}{1 - e^{-u}} e\left(-\frac{ru}{2\pi}\right) h(r) dr du$$

$$= \frac{1}{2\pi} \int_{\varepsilon'}^{\infty} \int_{-\infty}^{\infty} \left(e^{-\sqrt{-1}ru} + \frac{2e^{-u}}{u}\right) h(r) dr du$$

$$- \frac{1}{\pi} \int_{-\infty}^{\infty} \int_{\varepsilon'}^{\infty} \left(\frac{e^{-u}}{u} - \frac{e^{-u(1+\sqrt{-1}r)}}{1 - e^{-u}}\right) du\, h(r) dr$$

$$= \int_{\varepsilon'}^{\infty} g(u) du + 2g(0)\left(\log u \cdot e^{-u}\Big|_{\varepsilon'}^{\infty} + \int_{\varepsilon'}^{\infty} \log u \cdot e^{-u} du\right)$$

$$- \frac{1}{\pi} \int_{-\infty}^{\infty} \int_{\varepsilon'}^{\infty} \left(\frac{e^{-u}}{u} - \frac{e^{-u(1+\sqrt{-1}r)}}{1 - e^{-u}}\right) du\, h(r) dr$$

$$= -2g(0) \log \varepsilon' \cdot e^{-\varepsilon'} + \int_{\varepsilon'}^{\infty} g(u) du + 2g(0) \int_{\varepsilon'}^{\infty} \log u \cdot e^{-u} du$$

$$- \frac{1}{\pi} \int_{-\infty}^{\infty} \int_{\varepsilon'}^{\infty} \left(\frac{e^{-u}}{u} - \frac{e^{-u(1+\sqrt{-1}r)}}{1 - e^{-u}}\right) du\, h(r) dr.$$

Since ε and ε'^2 are almost equal when they are small, it follows from

$$\int_{0}^{\infty} \log u \cdot e^{-u} du = -\gamma, \qquad (\gamma = \text{Euler's constant}),$$

and from

$$\int_{0}^{\infty} \left(\frac{e^{-u}}{u} - \frac{e^{-u(1+\sqrt{-1}r)}}{1 - e^{-u}}\right) du = \frac{\Gamma'(1 + \sqrt{-1}r)}{\Gamma(1 + \sqrt{-1}r)}$$

that

$$\lim_{\varepsilon \to 0} \left(\log \varepsilon \cdot Q(\varepsilon) + \int_{\varepsilon}^{\infty} \frac{Q(w)}{w} dw\right)$$

$$= \frac{1}{2} h(0) - 2\gamma\, g(0) - \frac{1}{\pi} \int_{-\infty}^{\infty} \frac{\Gamma'(1 + \sqrt{-1}r)}{\Gamma(1 + \sqrt{-1}r)} h(r) dr.$$

Thus we have finally, for the second term of the right hand side of (A. 7),

$$\int_{1}^{\infty} f(x) dx = g(0) \log Y - (\log 2 + \gamma) g(0) + \frac{1}{4} h(0)$$

$$- \frac{1}{2\pi} \int_{-\infty}^{\infty} \frac{\Gamma'(1 + \sqrt{-1}r)}{\Gamma(1 + \sqrt{-1}r)} h(r) dr + o(1), \qquad (\text{as } Y \to \infty).$$

As for the third term of (A. 7), it follows from

$$f'(x) = -\frac{1}{x^2}\int_{x^2/Y^2}^{\infty}\frac{k(t)}{\sqrt{t}}\,dt - \frac{2}{Y}\frac{k\left(\frac{x^2}{Y^2}\right)}{x}$$

that

$$\int_0^{\infty} f'(x)\{x\}\,dx = g(0)\int_1^{\infty} -\frac{\{x\}}{x^2}\,dx + o(1).$$

On the other hand, it follows from $\frac{2n+1}{2}\left(\frac{1}{n+1}-\frac{1}{n}\right) = -\frac{1}{2}\left(\frac{1}{n}-\frac{1}{n+1}\right)$ that

$$\int_1^{\infty} -\frac{\{x\}}{x^2}\,dx = -\left\{\int_1^2 \frac{x-\frac{3}{2}}{x^2}\,dx + \int_2^3 \frac{x-\frac{5}{2}}{x^2}\,dx + \dots\right\}$$

$$= -\lim_{n\to\infty}\left\{\log(n+1) - \left(1+\frac{1}{2}+\dots+\frac{1}{n}+\frac{1}{2(n+1)}\right)+\frac{1}{2}\right\} = \gamma - \frac{1}{2}.$$

This implies that

$$\int_1^{\infty} f'(x)\{x\}\,dx = \left(\gamma - \frac{1}{2}\right)g(0) + o(1), \qquad (\text{as } Y\to\infty).$$

Summing up all these results related to (A. 7), we have for the first half of $c(\infty)$ in (A. 1),

$$\sum_{1\neq\gamma_0\in\Gamma_0}\int_0^Y\int_0^1 k(z,\gamma_0 z)dz = g(0)\log Y - \frac{1}{2\pi}\int_{-\infty}^{\infty} h(r)\frac{\Gamma'(1+\sqrt{-1}r)}{\Gamma(1+\sqrt{-1}r)}\,dr$$

$$- g(0)\log 2 + \frac{1}{4}h(0) + o(1), \qquad (\text{as } Y\to\infty).$$

The second half of $c(\infty)$ reduces to the evaluation of

$$\int_{\mathscr{D}_Y}\frac{1}{4\pi}\int_{-\infty}^{\infty} h(r)E(z,s)E(z,\bar{s})drdz$$

with $s = ½ + \sqrt{-1}r$, $(r \in \boldsymbol{R})$. For this purpose, we then use the inner product formula

$$(E^Y(z,s), E^Y(z,s)) = ||E^Y(z,s)||^2$$

$$= \frac{Y^{2S-1} - |\varphi(s)|^2 Y^{-2S+1}}{2S-1} + \frac{\varphi(\bar{s})Y^{2\sqrt{-1}r} - \varphi(s)Y^{-2\sqrt{-1}r}}{2\sqrt{-1}r}$$

in Theorem 2.3.2, valid for $s = S + \sqrt{-1}r$, $\operatorname{Re} s = S > ½$. The index i

is omitted because there is only one Eisenstein series in the present case. Since $h(r)$ is an even function, the functional equation $|\varphi(1/2 + \sqrt{-1}r)| = 1$ (Theorem 4.4.1) implies that

$$\lim_{S\to\frac{1}{2}} \int_{-\infty}^{\infty} h(r)\frac{Y^{2S-1} - |\varphi(s)|^2 Y^{-2S+1}}{2S-1}dr$$

$$= \lim_{S\to\frac{1}{2}} \int_{-\infty}^{\infty} h(r)\left(\frac{Y^{2S-1} - Y^{-2S+1}}{2S-1} + \frac{1-|\varphi(s)|^2}{2S-1}Y^{-2S+1}\right)dr$$

$$= 4\pi\, g(0)\log Y - \int_{-\infty}^{\infty} h(r)\frac{\varphi'\left(\frac{1}{2}+\sqrt{-1}r\right)}{\varphi\left(\frac{1}{2}+\sqrt{-1}r\right)}dr.$$

Furthermore

$$\int_{-\infty}^{\infty} h(r)\frac{\varphi(\bar{s})Y^{2\sqrt{-1}r} - \varphi(s)Y^{-2\sqrt{-1}r}}{2\sqrt{-1}r}dr = \int_{-\infty}^{\infty} h(r)\frac{\varphi\left(\frac{1}{2}-\sqrt{-1}r\right)}{\sqrt{-1}r}e^{2\sqrt{-1}r\log Y}dr$$

has the same limit as

$$\int_{-\infty}^{\infty} h(r)\frac{\operatorname{Re}\varphi\left(\frac{1}{2}-\sqrt{-1}r\right)}{r}\sin(2r\log Y)dr$$

as $Y\to\infty$, and the limit of the latter expression is equal to $\pi h(0)\varphi(1/2)$ because of the theorem

$$f(0) = \lim_{\lambda\to\infty}\frac{1}{\pi}\int_{-\infty}^{\infty} f(t)\frac{\sin\lambda t}{t}dt$$

on Fourier integrals. Here, some integrals which are not absolutely convergent should be interpreted as the Cauchy principal value, for instance. Thus,

$$\int_{\mathscr{D}_Y}\frac{1}{4\pi}\int_{-\infty}^{\infty} h(r)E(z,s)E(z',\bar{s})drdz$$

$$= g(0)\log Y - \frac{1}{4\pi}\int_{-\infty}^{\infty} h(r)\frac{\varphi'\left(\frac{1}{2}+\sqrt{-1}r\right)}{\varphi\left(\frac{1}{2}+\sqrt{-1}r\right)}dr + \frac{1}{4}h(0)\varphi\left(\frac{1}{2}\right) + o(1),$$

as $Y\to\infty$. This is the final formula for the second half of the expression for $c(\infty)$. Combining our results for the first and second halves of the expression for $c(\infty)$, we get

THEOREM A.1. The constant $c(\infty)$ in (A. 1) is equal to

$$\frac{1}{4\pi}\int_{-\infty}^{\infty} h(r)\frac{\varphi'\left(\frac{1}{2}+\sqrt{-1}r\right)}{\varphi\left(\frac{1}{2}+\sqrt{-1}r\right)}dr - \frac{1}{2\pi}\int_{-\infty}^{\infty} h(r)\frac{\Gamma'(1+\sqrt{-1}r)}{\Gamma(1+\sqrt{-1}r)}dr$$

$$- g(0)\log 2 + \frac{1}{4}\left(1-\varphi\left(\frac{1}{2}\right)\right)h(0).$$

Finally, let us take a look at the trace formula for a general discontinuous group Γ of finite type. The kernel function $K(z, z')$ is also defined in this case by $K(z, z') = \sum_{\sigma \in \Gamma} k(z, \sigma z')$, but $\hat{H}(z, z')$ should be defined as in (5.3.6). If $\kappa_1, \dots, \kappa_h$ form a complete set of inequivalent cusps of Γ, then $c(\infty)$ in (A. 2) should be replaced by

$$c(\infty) = \lim\left[\sum_{i=1}^{h}\int \sum_{1 \neq \gamma \in \Gamma_i} k(z, \gamma z)dz - \int \hat{H}(z, z)dz\right],$$

where, denoting by Γ_i the stabilizer of κ_i in Γ, and by σ_i a linear transformation such that $\sigma_i^{-1}\Gamma_i\sigma_i = \Gamma_0 = \left\{\begin{pmatrix}1 & b\\ & 1\end{pmatrix} \middle| b \in \mathbf{Z}\right\}$, we take the first integral over the domain consisting of all points z in the non-euclidean strip $\Gamma_i\backslash H$ such that $\operatorname{Im} \sigma_i^{-1} z < Y$, and the second integral over the domain consisting of all $z \in \mathscr{D} = \Gamma\backslash H$ such that $\operatorname{Im} \sigma_i^{-1} z < Y$ for all $i = 1, \dots, h$. The other components $c(I)$, $c(R)$ and $c(P)$ in (A. 2) need not be changed at all.

Using

$$(E_i^Y(z, s), E_i^Y(z, s))$$

$$= \frac{Y^{2S-1} - \sum_l |\varphi_{il}(s)|^2 Y^{-2S+1}}{2S-1} + \frac{\varphi_{ii}(\bar{s})Y^{2\sqrt{-1}t} - \varphi_{ii}(s)Y^{-2\sqrt{-1}t}}{2\sqrt{-1}t}$$

derived from Theorem 2.3.2, it is shown, essentially in the same way as in the case of $h = 1$, that the constant $c(\infty)$ for general Γ is equal to the trace of the $h \times h$ matrix obtained from the formula in the above theorem after replacing φ by the constant term matrix Φ and all other scalars by the diagonal obtained by multiplying them by the identity matrix of degree h.

Index